Management Technologies for Metal Mining Influenced Water

Geochemical Modeling for Mine Site Characterization and Remediation

Volume 4

Edited by D. Kirk Nordstrom and Andrew Nicholson

PUBLISHED BY THE
SOCIETY FOR MINING, METALLURGY & EXPLORATION

Society for Mining, Metallurgy & Exploration (SME)
12999 E. Adam Aircraft Circle
Englewood, Colorado, USA 80112
(303) 948-4200 / (800) 763-3132
www.smenet.org

The Society for Mining, Metallurgy & Exploration (SME) is a professional society whose more than 15,000 members represent all professionals serving the minerals industry in more than 100 countries. SME members include engineers, geologists, metallurgists, educators, students and researchers. SME advances the worldwide mining and underground construction community through information exchange and professional development.

ISBN 978-0-87335-353-3
eBook 978-0-87335-461-5

Library of Congress Cataloging-in-Publication Data

Names: Nordstrom, Darrell Kirk, editor. | Nicholson, Andrew (Environmental engineer), editor.
Title: Geochemical modeling for mine site characterization and remediation / edited by D. Kirk Nordstrom and Andrew Nicholson.
Description: Englewood, Colorado, USA : Society for Mining, Metallurgy & Exploration, 2017. | Series: Management technologies for metal mining influenced water ; volume 4 | Includes bibliographical references and index.
Identifiers: LCCN 2017037071 (print) | LCCN 2017037568 (ebook) | ISBN 9780873354615 | ISBN 9780873353533 | ISBN 9780873354615 (Ebook)
Subjects: LCSH: Abandoned mined lands reclamation. | Mine drainage. | Mine water. | Geochemical modeling.
Classification: LCC TD195.M5 (ebook) | LCC TD195.M5 G464 2017 (print) | DDC 622.028/6--dc23
LC record available at https://lccn.loc.gov/2017037071

DISCLAIMER

Contents

Editorial Board and Contributors

Senior Editors

D. Kirk Nordstrom, U.S. Geological Survey, Boulder, Colorado
Andrew Nicholson, Integral Consulting Inc., Louisville, Colorado

Contributors

Walter Weinig, Laramide Sciences, Lakewood, Colorado
Ulrich Mayer, University of British Columbia, Vancouver, British Columbia
Ann Maest, Buka Environmental, Boulder, Colorado

Preface

Geochemical Modeling for Mine Site Characterization and Remediation is the fourth of six volumes in the Management Technologies for Metal Mining Influenced Water series of handbooks about technologies for management of metal mine and metallurgical process drainage. This volume provides basic technical information on geochemical modeling as it pertains to mine site characterization and remediation for a wide range of parties, including mine planners and engineers, environmental managers, land managers, consultants, researchers, government regulators, nongovernment organizations, students, stakeholders, and anyone with an interest in mining influenced water. Although this handbook focuses on geochemical modeling, hydrologic modeling is included because mining contaminants most often migrate by surface water and groundwater transport, and contaminant concentrations are a function of water residence time as well as pathways.

This handbook makes important links to the other five handbooks. Geochemical modeling is used to describe the reactions and processes of acid mine drainage formation, secondary mineral formation, and carbonate neutralization outlined in Volume 1 (*Basics of Metal Mining Influenced Water*, McLemore 2008). Geochemical modeling can be applied to mitigation design and implementation described in Volume 2 (*Mitigation of Metal Mining Influenced Water*, Gusek and Figueroa 2009) because the chemical consequences of water diversion, treatment (e.g., lime neutralization), tailings, or waste rock removal can be estimated through modeling. Important themes in Volume 3 (*Mine Pit Lakes: Characteristics, Predictive Modeling, and Sustainability*, Castendyk and Eary 2009) include hydrogeochemical modeling of water quality predictions for pit lakes, which directly link to topics in this handbook. Geochemical modeling is often used to complement predictive testing from static and kinetic methods described in Volume 5 (*Techniques of Predicting Metal Mining Influenced Water*, Williams and Diehl 2014), and it often applies to the interpretation of water analyses that must meet quality assurance and quality control standards outlined in Volume 6 (*Sampling and Monitoring for the Mine Life Cycle*, McLemore et al. 2014).

These six handbooks are a volunteer project of the Acid Drainage Technology Initiative–Metal Mining Sector (ADTI-MMS). The Metal Mining Sector is part of the Acid Drainage Technology Initiative, which is the United States' regional member of the Global Alliance sponsored by the International Network for Acid Prevention. The mission of ADTI-MMS is to identify, evaluate, develop, and disseminate information about cost-effective and environmentally sound methods and technologies to manage mine wastes and related metallurgical materials for abandoned, inactive, active, and future mining and associated operations, and to promote understanding of these technologies.

This volume greatly benefitted from reviews by several experts. We especially thank Madhumita Raghav, Eric Weiland, Rens Verburg, Roderick (David) Williams, and Charles Bucknam for their reviews and their patience during the preparation of this volume. Gratitude and thanks are extended to Zoë Polando for her renderings of several figures in Chapter 6. The senior editor and author (DKN) thank the National Research Program of the U.S. Geological Survey for their support, without which this volume could not have been completed.

CHAPTER 1

Introduction

D. Kirk Nordstrom

OBJECTIVE

The objective of this handbook is to describe the important components of hydrogeochemical modeling for mine environments, primarily those mines where sulfide minerals are present—metal mines and coal mines. This chapter provides a general introduction to the overall topic of geochemical modeling and mine wastes. Because the reliability of modeling has often been questioned on philosophical and factual grounds, a clarification of terms was deemed important. Consequently, a discussion on the strengths and limitations of modeling is included in Chapter 2. Geochemical modeling, like any type of modeling, is meaningful only within a defined system that identifies boundary and initial conditions and utilizes mass balances (Chapter 3) within that system. From a regulatory perspective, modeling is commonplace and sometimes helpful, yet sometimes counterproductive. Chapter 4 suggests important aspects that should be considered during regulatory review of modeling results. The major hydrogeochemical processes occurring in mine wastes are discussed in Chapter 5. These processes include pyrite oxidation, mixed pyrite and carbonate dissolution, gypsum dissolution and precipitation, oxidation–reduction properties of trace metals and metalloids, iron precipitation, and aluminum dissolution and precipitation. Chapter 6 covers the fundamental aspects of hydrologic modeling, while Chapter 7 reviews reactive transport modeling with an example and a discussion of its strengths and limitations. Database quality is critical input to geochemical models, and Chapter 8 provides a brief description of popular codes used to model geochemistry for mine sites, along with some discussion of test cases and examples given by some codes, code benchmarks, and an introduction to thermodynamic databases. Chapter 9 describes a variety of case studies that show how geochemical models have been applied to characterization or remediation of mine wastes.

The appendixes are included to supplement the main text. Appendix A provides a recapitulation of modeling limitations. Appendix B is the senior author's perspective on a 1996 paper by Glynn and Brown. A reference list is provided at the end of the handbook to identify more detailed information on topics of interest.

MODELING AND MINE WASTES

Models are used extensively for numerous technical, scientific, engineering, and regulatory purposes. They have a long history of use as tools to help us understand and explain physical, biological, and chemical phenomena and to predict outcomes. They have also been misused. Models have been equated with computer codes (blatant misuse of language), and code output has occasionally replaced expert judgment or even common sense. This handbook provides some general guidelines on the strengths and limitations of geochemical modeling and an overview of its application to the hydrogeochemistry of both unmined mineralized sites and those contaminated from mineral extraction and mineral processing.

Opinions on modeling vary from "useless as a quantitative exercise for policy matters" (Pilkey and Pilkey-Jarvis 2007) to "an essential and powerful tool for understanding natural water–rock systems" (Zhu and Anderson 2002). The authors of this handbook share most of these opinions, and they are not necessarily incompatible. The single most-important factor to the successful application of a geochemical model is the knowledge and experience of the individual or group conducting the modeling. An experienced hydrogeochemist can anticipate potential water quality trends and issues at a mine site even before applying a computer code because he or she has studied similar sites and has a working knowledge of what the important factors are. His or her conceptual model is fine-tuned based on the experience that the modeler brings to the project.

Mine environments consist of complex geologic, hydrologic, and biologic processes containing both natural and engineered features. One cannot hope to understand all the complexities that give rise to a particular water quality, air quality, soil quality, or sediment quality, so some assumptions must be made in concert with good quality data and deterministic principles that are understood. These concepts along with necessary assumptions and available data can be coded in a computer program that can execute thousands of calculations in milliseconds. The results, or output, allow one to examine numerous hypotheses for geochemical processes that could explain the observations and to anticipate future trends. This output is only useful in the hands of those who are trained and experienced in hydrogeochemical processes and in geochemical modeling. This handbook cannot provide that training or experience; it can only give an overview of the models behind the codes, explain the geochemical computations, describe some of the modeling processes, provide a compilation of codes, and give examples of their application, including both successes and failures. The intention is to keep the material at a general level without too much technical or theoretical detail but with sufficient information to be of practical use.

"Models" have been so often abused, misused, and overused that they have, in some instances, replaced clear thinking. Consequently, this handbook begins with definitions of terms, a discussion of the role of models, what purpose they can serve, whether they can predict anything, and some general guidelines for their use. The most important models are conceptual models, that is, one's concepts of physical, biological, or chemical phenomena. When models do not work, when codes give outlandish output, and when decisions are made based on such unreliable sources, it is usually because the conceptual models are incorrect.

Indeed, it is so easy (and inexpensive) to produce computer output with canned or new programs that one often loses sight of what the purpose and limitations of models are. Having gone so far down this path, it is common to make the mistake of calling a computer code a model. It has often been said that models are no better than their input data (input data not only apply to computer models but also to noncomputer calculations done with a calculator, or even by hand if one remembers how). Indeed, the models can be even worse than their input data. Realistic constraints on water–rock interactions come from observations of the geology, hydrology, and chemistry of a site that one is trying to understand, not from a model or a code. The more field data and the better quality the field data, the easier it becomes to develop a feasible model that portrays a useful and reliable understanding of a complex process.

Hydrogeochemical processes take place in the four dimensions of space and time (x, y, z, t). Our concepts and modeling must also apply within this context. Water, rock, and biota move and evolve not only in complex patterns but they do so within a spatial and temporal context. Spatial and temporal changes increase the variables and degrees of freedom, but with additional physical and chemical observations, enough constraints can usually delineate the main hydrobiogeochemical processes at a given field site.

CHAPTER 2

What Is a Model?

D. Kirk Nordstrom

INTRODUCTION

A model is a simplification of reality. It is often said that a model is a representation of reality, but it would be more correct to say that a model represents our thinking about reality. Reality is too complex to be understood without considerable simplification. Every time we generalize and communicate our experiences, our feelings, our observations, our deductions, and our inferences, we are forced to simplify and abstract them into forms or concepts that can be communicated. This simplifying abstraction is necessary for understanding complex phenomena and for communicating our understanding. It is part of the human condition. All human endeavors—arts and sciences, religion and philosophy, politics and economics—are subject to abstraction and simplification. The distinguishing feature of scientific models is that they are a simplification of reality. It is physical, biological, or chemical phenomena we are trying to understand. A model gives us a "picture" of how something works without too many details, uncertain assumptions, and unnecessary conjecture.

The awkward aspect of simplification is the possibility of missing a key process or property that is needed to understand a site under investigation. Consequently, one should not rely on modeling alone, but include checks on modeling results such as expert judgment, peer review, high-quality field data, reliable test cases, and testing of modeling conclusions with new, independent observations.

SCIENTIFIC MODELS

Scientific models begin as ideas and are formalized into a language that is often, but not necessarily, mathematical. A *scientific model* is a testable idea, hypothesis, theory, combination of theories, or the context of theories that provides new insight or a new interpretation of an existing problem or question (Nordstrom and Campbell 2014). An often-quoted attribute is testability (Popper 1934); another is logical consistency; another is coherence with known facts; and another is simplicity. Powerful simplicity (i.e., a minimal number of concepts, principles, and equations that explain a maximum number of observations, or what is known as Ockham's razor), logic, and reproducibility by others have long been hallmarks of the scientific method. *Testability* means it has the possibility of being tested, and *reproducibility* means it can be tested and confirmed by others. Greenwood (1989) stated it well when he said that the word *model* should be reserved "... for well-constrained logical propositions, not necessarily mathematical, that have necessary and testable consequences, and avoid the use of the word if we are merely constructing a scenario of possibilities." Unfortunately, "modeling" is not just applied to testable constructs, but often to untestable scenario constructs as well. If it is untestable, it is not helpful science and it may not be meaningful at all.

Types of Scientific Models

Models come in different forms, usually dictated by the objectives of research. *Conceptual models* are the most fundamental and are described in the following section. For example, everyone has some concept of water–mineral interactions and the water flow regime. Consider a groundwater reacting with aquifer minerals. During its evolution, one might conceive some minerals to be undersaturated in the area of recharge, but other minerals (those that dissolve fastest) would become saturated somewhere downgradient, having reached their equilibrium solubility (a state of partial equilibrium). The conceptual model can be formalized into mathematics, the *mathematical model*, using chemical principles and entered into a computer program, the *code*, and predictions made to test the assumptions (and the databases) against the results from real field data. This exercise helps to quantify and constrain the possible reactions that might occur in the subsurface. Mathematical equations for complex interacting variables are not always solved exactly, and, therefore, systems of numerical approximations, or *numerical models*, have been developed. Alternatively, an experiment could be set up in the laboratory with columns made of aquifer material with groundwater flowing through to simulate reactive transport, the *experimental model* or *scale model*. Having obtained results from a mathematical or scale model, some unexpected results often occur that force us to change our original conceptual model. This example demonstrates how science works: it is an ongoing process of approximation that iterates between idea, theory, observational testing of hypothesis and theory, and back to modifications of theory or development of new theories.

Conceptual models. The importance of the conceptual model cannot be understated. As Bredehoeft (2005) has pointed out, "Every model has at its foundation a *conceptual model*." Because of the ultimate importance of the conceptual model, it is worthwhile to elaborate. For example, how does the conceptual model originate? There are no simple, satisfying answers to this question; it has been debated by philosophers and scientists for centuries. It arises from the conventional knowledge available at the time, the extent to which the modeler has acquired that knowledge, and the creative capability of the modeler. It is a combination of acquired knowledge and creative inference. Bredehoeft (2005) reminds us of the following:

- *The conceptual model is based on the subjective judgment of the analyst* [To which we would add that this judgment is objectively constrained by experience].
- *A numerical model provides a tool by which to test the appropriateness of the prevailing concept.*
- *One can expect the conceptual model to be continuously updated as new information is acquired.*

Serious errors in modeling most often arise from incorrect concepts and/or incorrect assumptions that most strongly affect the model results. Accordingly, someone untrained in basic concepts of geology, hydrology, and geochemistry should not be attempting to model geochemical processes. Bredehoeft (2003, 2005) also warns:

- *Modelers tend to regard their conceptual models as immutable.*
- *Time and again, errors in prediction revolve around a poor choice of the conceptual model.*
- *More often than not, data will fit more than one conceptual model equally well.*
- *Good calibration of a model does not ensure a correct conceptual model.* [This point refers to groundwater flow models because geochemical models are not usually calibrated. However, it can be said that a good match between predicted and

observed geochemical parameters does not ensure a correct conceptual model. It is not even a necessary condition, let alone a sufficient one.]

- [paraphrased] *Sensitivity analysis, probabilistic sampling, or error propagation analysis cannot compensate for uncertainties in the conceptual model, or for wrong or incomplete models, or for errors in the input data.*

These warning messages are extremely important and, if not understood and followed, can lead to major mistakes in any modeling effort.

CONCEPTUAL ERRORS WITH MODELS AND MODEL VALIDITY

Much has been written about model validation (Konikow and Bredehoeft 1992; Oreskes et al. 1994; Anderson and Bates 2001; Nordstrom 2012; and references therein) and there is considerable confusion about exactly what it means. It is partly a semantic issue, but it goes deeper into the question of what science and engineering are, because part of the disagreement is along the lines drawn between these two disciplines. How do we demonstrate that our models can calculate future scenarios reliably? The simple answer is that scientific models are found reliable only by testing with observations in the future. However, future scenarios can be hundreds to hundreds of thousands of years into the future. Such futuristic modeling would seem not to be rational because of the lack of ability to test consequences. But there are even more fundamental questions that should be noted.

Does good agreement between model results and observational measurements mean a model is correct? No, for three basic reasons: (1) if model parameters are not independent from the measurements they are being compared to, they should agree and we learn little from the model; (2) if the measurements are in error, then both the measurements and the model could be in error or the model could be correct by coincidence; and (3) the model results might agree with good measurements for the wrong reasons.

Does poor agreement between model result and observation mean the model is incorrect? No, for similar reasons: (1) if the measurements are unreliable, the model may still be correct; (2) model calculations could be in error, whereas the conceptual model could be correct; and (3) the criteria for good and poor agreement may be incompatible with the limitations and uncertainties of the model (i.e., the model results might actually agree with measurements, but the error window on what constitutes good agreement is unrealistically narrow).

For some situations, it has been demonstrated that the modeling was fallacious, usually because the conceptual model was inadequate or erroneous. Does this mean that the modeling was useless, meaningless, or a waste of time? Not necessarily. Sometimes an important new phenomena is discovered even with an incorrect model (Hodges 1991). It does, however, mean that modeling is not always useful and that more data are usually necessary to improve the conceptual model and adequately constrain any quantitative modeling.

Before the idea is conveyed that it might be useless to compare model results with independent observational data, it must be clarified that it is useful and always will be an essential aspect of the scientific method. Theoretical calculations must always be compared with observations. However, it is just as important to realize that a good comparison is not a guarantee that a model is "valid" or correct or even meaningful. A number of possible factors can cloud the interpretation of such comparisons. As more data become available, more calculations are made, a deeper understanding of the site properties and processes is gained, and a closer approximation of reality is achieved, but it is always an approximation. Consequently, not only must an experienced geochemist (not just a modeler) use his or her practical knowledge to interpret computational output, but independent peer review should also pass judgment on the results. Peer review is an essential and standard procedure of the scientific method to increase confidence in the conclusions and to reject poorly or incorrectly interpreted conclusions.

It is best to think of models as helpful guides to improve our understanding, as heuristic tools to organize our thoughts, as useful constraints on our conceptual thinking, and, when embodied in a computer code, as a means of handling large, complex sets of data that can be tested to limit and constrain hypothetical possibilities, not to prove some future outcome.

BACKGROUND PREREQUISITES FOR MODELING

The successful application of geochemical models assumes familiarity with, at a minimum, geology, hydrology, geochemistry, and microbiology. One of the most useful aspects of geology is mineralogy, especially mineral identification, their chemical compositions, and relative solubilities and stabilities. An understanding of rock types or lithologies, their main minerals, and their mode of formation are equally important. In hydrology, knowledge of water balances and the water cycle, Darcy's law, the principle of continuity, and the relation of geology to groundwater flow are essential. More than a dozen basic processes should be understood to properly describe the typical geochemical attributes of a mine environment. These processes include aluminosilicate weathering and clay formation (kaolinitization and illitization), carbonate weathering, dolomitization, dedolomitization, evaporation and evaporite mineral formation, ion exchange, gas evolution (in-gassing and de-gassing), geothermal and metasomatic alteration, mixing of chemically different waters, and processes related to microbiology, such as sulfide-mineral oxidation, iron and manganese reduction and oxidation, nitrate reduction and ammonium oxidation, sulfate reduction, methanogenesis, and hydrogenogenesis. Each of these processes has chemical and isotopic signatures associated with it and each might dictate different approaches to sampling and interpretation. This handbook will not address all of these processes, only those that are most relevant to mine environments.

Within the context of solute mobilization in mine environments, the following geochemical processes are considered most important:

- Sulfide-mineral oxidation
- Carbonate mineral dissolution and precipitation
- Gypsum dissolution and formation
- Aqueous oxidation and reduction of trace metals and metalloids (such as Fe, Mn, As, S, Se, Sb, Cr, U, and V)
- Iron mineral precipitation
- Aluminosilicate dissolution and precipitation
- Aluminum precipitation
- Silica precipitation
- Efflorescent salt formation and dissolution
- Mixing of waters with reaction
- Evaporation of waters with reaction

Each of these processes is briefly described in Chapter 5 in concert with the application of models that either interpret or simulate the processes. More details can be found in textbooks by Langmuir (1997) and Appelo and Postma (2007).

CHAPTER 3

Geochemical Modeling and Environmental Systems

D. Kirk Nordstrom

INTRODUCTION

Geochemical models utilize concepts from physical chemistry, that is, from thermodynamics, kinetics, and quantum mechanics. Although impressive contributions of quantum mechanics have been made to our understanding of hydrogeochemistry, this chapter only considers a few basic concepts from thermodynamics, kinetics, and mass balances. These topics have proven most useful to the application of models for site characterization and remediation.

Hydrogeochemical models can be static (time-independent) or dynamic (time-dependent). They apply to some defined part of our environment known as *the system*. The system is defined by the modeler but should include spatial and temporal boundaries. For example, waste rock or tailings have physical dimensions of length, width, height, mass, and volume that define them. These physical dimensions can change over time because of a variety of processes including erosion, renewed mining, and remediation.

A system may be open, closed, adiabatic, or isolated. An *open system* allows the exchange of both matter and energy across its boundaries. Waste rock, if not sequestered or capped, is usually permeable to the transport of air and water through it and is typically an open system. Waste rock is subject to heating by the sun during the day and cooling by night as well as cooling from rainwater infiltration. Also, it can produce heat from pyrite oxidation and release that heat to its surroundings which then affects the transport of air and water through convective cycling. Some tailings are managed by keeping a permanent water cover over them so that they are closed to air transport but open to water transport. They are clearly open systems if not contained in some way. A *closed system* allows energy, but not matter, to move across the boundaries. Waste rock with an impermeable liner at the base and impermeable liners on the sides and top would be an example of a closed system. It would still be affected by outside temperature gradients from solar cycles and changing weather conditions, and if it was producing heat, that heat could be lost to the surrounding air and land. An *adiabatic system* is a closed system that allows work energy but no heat energy to cross the system boundaries. A good example of an adiabatic system is a geothermal fluid rising in a silica-lined fracture to discharge as a geyser or hot spring. It rises because of pressure gradients and can expend work energy but often does not transfer much heat with the surrounding rock (which may be at the same temperature). An *isolated system* is completely cut off from any flow of energy or mass across its boundaries.

Environmental systems usually involve mass transport such as water flow and solute flow. Transport can fall into one of three categories according to their time-dependency: equilibrium state, steady state, and transient state.

The Equilibrium State

The proper definition of an *equilibrium state* is an isolated system in which there are no gradients in energy or mass. All measurable parameters are constant throughout the dimensions of the system (space-invariant) and for the period of time of the measurements (time-invariant). By the very definition of being isolated, such a system will reach an equilibrium state given sufficient time for all gradients to disappear and all parameters to equalize. A water table in an aquifer, if completely isolated from all inflows and outflows and other driving forces, will become horizontal if only acted on by the force of gravity.

In practice, the equilibrium state is not a typical situation because of all the dynamic forces that operate within the environment. It could be said that given geologic periods of time or just realistic definitions of any system, the equilibrium state is never reached. What, then, is the usefulness of an equilibrium state? There is tremendous usefulness in the concept for several reasons. First, it is a practical reference point for a system. If it is known what defines equilibrium, then a determination can be made as to how close or far the system is from equilibrium. The simple example of the water table seems obvious. The greater the hydraulic gradient, the faster the tendency for the water table to return to an equilibrium state. A less obvious example is mineral dissolution; the further a mineral is from solubility equilibrium, the faster it will tend to dissolve to reach equilibrium. Second, although systems tend not to reach overall equilibrium, parts of the system do reach equilibrium, the *partial equilibrium state*, and sufficiently small volumes of the overall system can also reach equilibrium, the *local equilibrium state*. For instance, a groundwater might reach equilibrium quickly with respect to gypsum solubility, less quickly with respect to calcite solubility, and not at all with respect to quartz solubility because of the difference in dissolution rates. The overall system may be in equilibrium with gypsum, there may be limited areas in equilibrium with calcite, and equilibrium with quartz may only be present at the near-molecular level adjacent to quartz grains. Another example would be reactions involving aqueous species only. Most aqueous complex or ion-pair formation occurs rapidly (seconds or less) relative to groundwater flow and rapidly achieves equilibrium so that equilibrium-constant expressions can satisfactorily describe their chemistry. Third, although equilibrium is an essential concept in thermodynamics, it is also a powerful tool when combined with mass balances for the interpretation of water–rock interactions. Fourth, equilibrium thermodynamics is used in kinetic equations to describe rates of reaction.

The Steady State

When a system has temperature, pressure, or chemical gradients within its boundaries, its parameters will vary spatially (space-variant), but if those parameters do not change at any one location with time, the system has achieved a steady state. A time-invariant and space-variant system is the definition of a *steady state*. It can be an open or a closed system, but it cannot be isolated. Matter and/or energy must continue to flow in and out of the system to maintain parameter gradients. If recharge to and discharge from a groundwater aquifer are constant, the groundwater will move and maintain a constant head gradient. Steady state is often an approximation to the actual system being studied, not a true representation. This representation, however, can be an eminently useful and practical approach for modeling purposes.

The Transient State

Hydrogeochemical processes are technically in transient conditions. Groundwater flow is often treated as a steady-state system because it is a good approximation and the model setup and calibration are easier. A transient state, or space- and time-variant system, requires much more input data to calibrate even though, conceptually, it can be a more realistic condition. It is only for a limited range in space and time that groundwater flow, mineral dissolution or precipitation, and gas evolution can be approximated by the equilibrium state or the steady state. All surface waters and aquifers respond to short-term changes in seasons and long-term changes in climate by changing their water levels and flow rates. Aquifers will also change water levels and flow rates when the permeability of the aquifer changes. The transient state can be subdivided further according to thresholds for phenomena such as laminar versus turbulent flow, shear force of water across mineral surfaces that would allow or not allow equilibrium solubility to be reached, and degree of mineral supersaturation that would allow or not allow precipitation to occur at a rate sufficient to affect the water composition. Transient conditions, although the rule, are more difficult to model mathematically. Perhaps more importantly, considerable time and expense are required to obtain the necessary field input data and other parameters for modeling.

REACTION RATES AND WATER-FLOW RATES

The relationship between water-flow rate and mineral precipitation/dissolution rate is important and determines whether mineral solubility equilibrium is possible or not. At sufficiently high flow rates, solubility equilibrium is not possible, not even for very soluble minerals such as halite and gypsum because they require finite residence times in the same water volume to reach saturation. These soluble minerals will dissolve, but their dissolution rates determine the water composition, not equilibrium solubility. At sufficiently low flow rates, even under stagnant no-flow conditions, soluble minerals will reach equilibrium quickly, but not insoluble minerals. Nonequilibrium may be caused by different factors such as kinetic inhibition, mineral instability under ambient conditions, degree of crystallinity, surface area, and coatings of secondary minerals that can decrease or prevent equilibrium. Diamond is so insoluble and inert to aqueous solution that it cannot dissolve. Minerals such as quartz and zircon can dissolve, but they do so with great difficulty and require geologically long residence times in the same water, which is not a common situation. Generally, the more insoluble the mineral, the longer it takes to dissolve and reach equilibrium solubility. Also, many relatively insoluble minerals, such as jarosite, illite, and kaolinite, tend to be those with more complex formulae. In general, the more complex the formula, the more difficult it will be for the mineral to precipitate at low temperatures (<100°C [212°F]). Many soluble minerals have relatively simple formulae, such as halite and gypsum. Quartz, diamond, and zircon are obvious exceptions because they have simple formulae and yet are very insoluble. The governing factor is the type of chemical bonding in the crystal structure and its tendency to interact with the polar bonds of water molecules. Quartz, diamond, and zircon have strong covalent bonds in a continuous three-dimensional network that have no affinity for ionized water molecules (Pauling 1964).

Hence, there are three key rates that govern observed water compositions: (1) the rate of mineral dissolution, (2) the rate of mineral precipitation, and (3) the flow rate of the groundwater or surface water. If mineral dissolution and precipitation rates are fast relative to the water-flow rate, equilibrium solubility can be achieved more easily. For the reverse situation, equilibrium is unlikely to be achieved. Although many dissolution rates of common minerals have been measured, they have not been determined for the entire range of water composition and temperature found in all groundwaters. Furthermore, lab studies of mineral dissolution rates often do not compare well with field studies. The most realistic rates come from direct measurements of water residence times combined with mass balance calculations (e.g., Burns et al. 2003). It is important to know relative rates of mineral dissolution and precipitation to properly vet the output from geochemical computations. Mineral dissolution and precipitation kinetics is still an active area of research.

MASS BALANCES

Basic Principles and Mixing Relations

One of the most basic laws of science is the law of conservation of mass. It is the very first chapter in Albarède's (1995) textbook, *Introduction to Geochemical Modeling*. The law is central to understanding and confirming chemical reactions, microbiological processes, water flow in drainage basins, water–rock interactions, and solute-transport processes. It is a powerful accounting tool that should be the core of interpreting any geochemical process by quantitative methods. The law is simple in concept, but not always simple in practice. It can be used in static calculations, such as relating water chemistry to mineral and gaseous reactions, and in dynamic calculations, such as mixing of waters with different compositions.

In the simplest formulation, we can consider the mixing of two water masses, m_1 and m_2:

$$m_1 + m_2 = m_3 \qquad \text{(EQ 3.1)}$$

These can be converted to volumes, V, by dividing through by a constant density, ρ, to obtain

$$V_1 + V_2 = V_3 \qquad \text{(EQ 3.2)}$$

If we consider water flow rates, that is, $V/t = Q$ (or discharge), where t is time, we have

$$Q_1 + Q_2 = Q_3 \qquad \text{(EQ 3.3)}$$

Equation 3.3 is one of two equations that can be used to solve for concentrations of a contaminant after mixing of two streams of water with different compositions. Considering an ion, i, at a concentration of C_1^i in water mass 1 mixing with water mass 2 at a concentration of C_2^i, we can determine the concentration after mixing by conservation of mass:

$$C_1^i Q_1 + C_2^i Q_2 = C_3^i Q_3 \qquad \text{(EQ 3.4)}$$

because CQ is the mass flux of a solute (or m/t obtained by just dividing through Equation 3.1 by t). Equation 3.4 assumes no chemical reaction (i.e., these ions are behaving conservatively in solution). As long as we know any two of the three discharges and any two of the three concentrations, the third concentration and discharge can be calculated. These equations are simple forms of the *continuity equation* that is fundamental to all fluid dynamics, including heat, electricity, ions or chemical components, and particles (e.g., Bird et al. 1960). *Continuity* refers simply to the transport of a conserved quantity.

Equation 3.4 can also be used to define a mixing line between two waters of distinctly different composition, even when the discharges are not known. If Equation 3.3 is divided by Q_3 such that $Q_2/Q_3 = 1 - Q_1/Q_3$, then by substitution into Equation 3.4 we have

$$\frac{C_1^i Q_1}{Q_3} + C_2^i \left(1 - \frac{Q_1}{Q_3}\right) = C_3^i \qquad \text{(EQ 3.5)}$$

And given that Q_1/Q_3 is just a mixing ratio, we can call it x and then

$$C_1^i x + C_2^i (1-x) = C_3^i \qquad \text{(EQ 3.6)}$$

If one water type is 100% of mass 1, the concentration of the ith ion is defined by C_1^i, and if 100% mass 2, then its concentration is defined by C_2^i. Anything in between is defined by the mixing ratio. Imagine mine water seepage with a sulfate concentration of 500 mg/L entering a creek with a sulfate concentration of 20 mg/L before mixing. If, after mixing, the sulfate concentration is 50 mg/L, what is the proportion of mine water seepage relative to creek water flow? Solving for x we find that the mine seepage discharge is close to 6% of the flow of the creek. Alternatively, if we know the flow of the creek or the seepage water, we can determine what the final sulfate concentration is after mixing.

Equation 3.6 can be rearranged to a common linear form:

$$\left(C_1^i - C_2^i\right)x + C_2^i = C_3^i \qquad \text{(EQ 3.7)}$$

which results in a linear plot when one element concentration is plotted against another, and the slope is determined by the difference in concentration for the end-member waters, and the intercept is the concentration of one of the end-member waters.

Some concentration ratios are distinctly different between waters that mix, and these ratios can be assumed constant for each water type involved in the mixing process. Ratios allow us to distinguish what water compositions are derived from just simple dilution and which are derived from reaction in addition to dilution. Ratios also provide more information in a simple two-dimensional plot than could be portrayed otherwise. Dividing by another solute concentration, C^j, we have

$$\frac{\left(C_1^i - C_2^i\right)x + C_2^i}{\left(C_1^j - C_2^j\right)x + C_2^j} = \frac{C_3^i}{C_3^j} \qquad \text{(EQ 3.8)}$$

which forms an exponential-like curve when plotting the ratio against mixing fraction (Figure 3.1a).

An alternate way of portraying mixing is to plot the ion ratio relative to an ion concentration that occurs in the ratio. For example, C_3^i/C_3^j could be plotted against C_3^j, which results in a curved line (Figure 3.1b) or against $1/C^j$, which gives a straight line, and any point along the line represents the proportion of mixing (Figure 3.2).

A simple example demonstrates how mixing can be represented by these equations. Consider an uncontaminated mountain stream with 0.05 mg/L of manganese and 40 mg/L of sulfate, giving a mass ratio of 0.00125. Into this stream, some acid mine water with a manganese concentration of 35 mg/L and a sulfate concentration of 2,100 mg/L mixes (mass ratio of 0.0167). A plot of Mn/SO_4 mass ratios against the mixing proportion is shown in Figure 3.1. A major shift can be seen in the ratio when only 5% of acid mine water has mixed with the stream because the concentrations are so high in the mine water relative to the stream water. Because of the analytical ability to determine Mn and SO_4 very reliably, the water chemistry can be a sensitive indicator of mixing fractions.

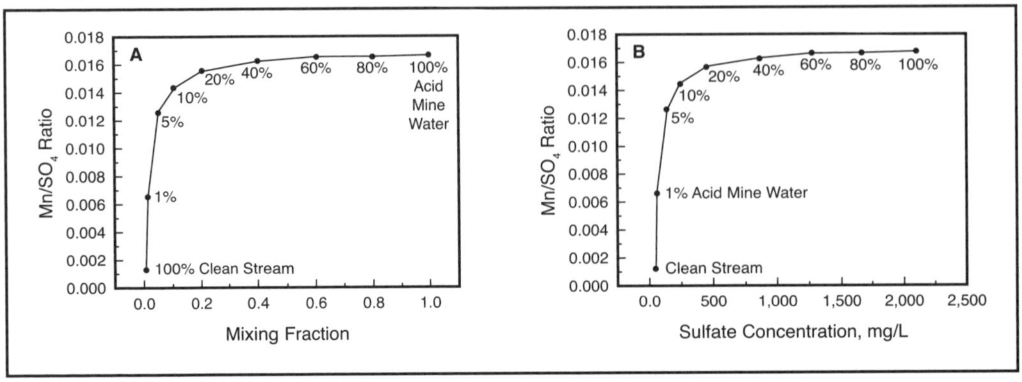

Courtesy of U.S. Geological Survey

FIGURE 3.1 (a) Hypothetical mixing between an acid mine water with a high Mn/SO_4 ratio and clean stream with a low ratio; (b) same mixing plot as in Figure (a) but changing the x-axis to sulfate concentration

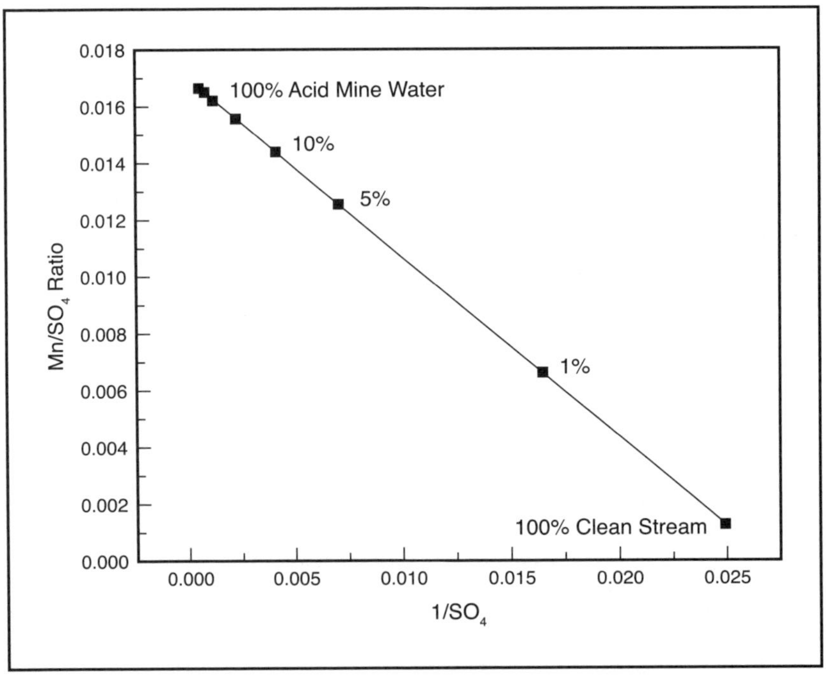

Courtesy of U.S. Geological Survey

FIGURE 3.2 Linearized plot of same data in Figure 3.1 using the inverse of the sulfate concentration for the x-axis

These plots can be linearized by using the inverse of the sulfate concentration (or mixing fraction) for the x-axis. A linearized plot also has the advantage of expanding the dilute end of the scale where one most commonly has data. Figure 3.2 is the same data from the previous two plots using this transformed axis.

Mixing of different waters is a very common phenomenon for both surface waters and groundwaters. Whenever mixing is part of the geochemical modeling, these principles must be followed for conservative species. When chemical reactions are involved, the modeler needs to find a conservative constituent (usually chloride or sulfate in oxidized systems) that can serve as a reference point for mixing without reaction.

Input-Output Mass Balances (Flux-Reservoir Modeling)

The continuity equation is often expressed as an input and output to a given volume element (the system):

$$\begin{Bmatrix} \text{rate of} \\ \text{mass input} \end{Bmatrix} - \begin{Bmatrix} \text{rate of} \\ \text{mass output} \end{Bmatrix} = \begin{Bmatrix} \text{rate of} \\ \text{mass accumulation/depletion} \end{Bmatrix} \quad \text{(EQ 3.9)}$$

or input – output = change in storage (S). If there is no accumulation or depletion, there would be a steady-state system in which

$$\begin{Bmatrix} \text{rate of} \\ \text{mass input} \end{Bmatrix} = \begin{Bmatrix} \text{rate of} \\ \text{mass output} \end{Bmatrix} \quad \text{(EQ 3.10)}$$

For a surface water balance of a catchment or a drainage basin, input = precipitation (P), output = evapotranspiration (ET) and infiltration (I) and runoff (R) (Freeze and Cherry 1984; Hornberger et al. 1998). Change in storage, ΔS, could refer to a lake or reservoir or to a groundwater aquifer:

$$P - ET - I - R = \Delta S \quad \text{(EQ 3.11)}$$

where runoff might include surface water inflow and infiltration, surface water outflow, and might include groundwater springs or other discharges to surface waters. Under steady-state conditions, this would reduce to

$$P = ET + I + R \quad \text{(EQ 3.12)}$$

Flux-reservoir modeling. Another type of input-output model is better known as flux-reservoir modeling because it consists of depicting large-scale systems, such as mine sites or even global cycles, as a set of reservoirs containing known amounts of a substance with rates of transfer between these reservoirs with (hopefully) known fluxes. A good example is the global water budget, because it applies Equation 3.12 to account for the total mass of water on and in the earth and the transport of water between these compartments. The total mass of water for the earth is approximately 1,459 × 10^{18} kg, which is divided between water vapor and liquid water, as shown inside the boxes in Figure 3.3. Approximately 96% of the world's water comprises the oceans, while freshwater resources (lakes, rivers, groundwaters) are only 1% of the total. Water is constantly in motion, transported from one reservoir to another. The average residence time for water in the atmosphere is only about 11 days because of the rapid rate of evaporation and precipitation.

A variation of flux-reservoir models is known as "stocks-and-flows" models associated with system dynamics (Ford 1999). As an example, Ford developed flow diagrams for Mono Lake in California, which begins with a simple model shown in Figure 3.4. The double-lined arrows represent mass flow, in this system water transport. Open circles are converters and single arrows are connectors. Connectors show the flow of information in a system. Convertors without any arrow going to it would be considered model input.

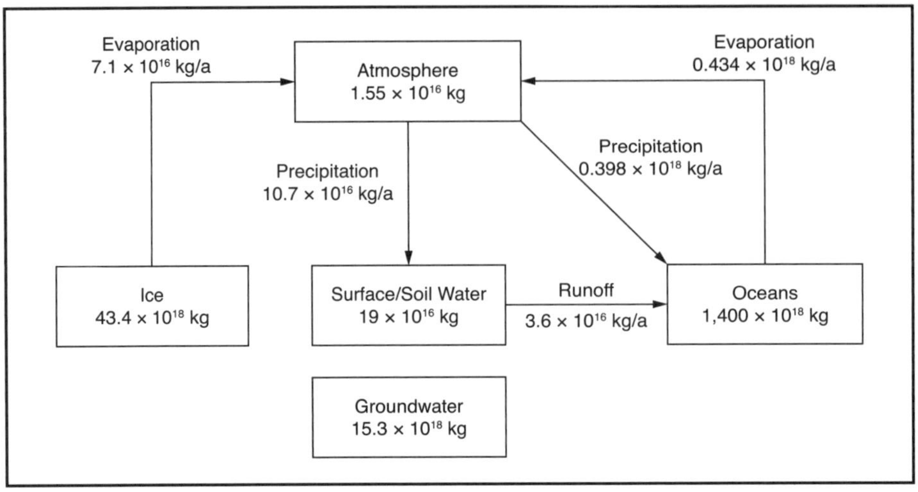

Data from Berner and Berner 1996

FIGURE 3.3 Global water budget

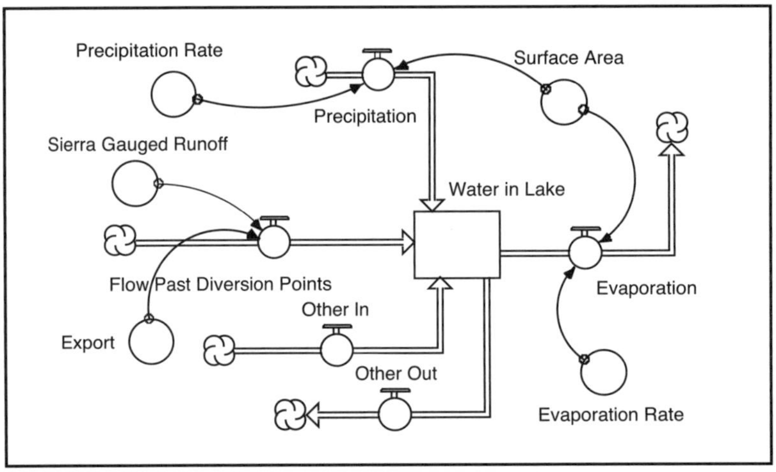

Source: From *Modeling the Environment*, 2nd edition, by Andrew Ford. © 2009 by Island Press. Reproduced with permission of Island Press, Washington, D.C.

FIGURE 3.4 First model for Mono Lake

A simple flux-reservoir model for the Iron Mountain Superfund mine site in Northern California (Alpers et al. 2003) can be portrayed, but first a schematic diagram is needed to better understand the site features, as shown in Figure 3.5. In this situation, the reservoir is a solid phase, pyrite, existing as massive lenses and as disseminated grains in rhyolite, hydrothermally altered by seawater. These large quantities of pyrite are reacting with oxygen and meteoric water, infiltrating the site to produce effluent with extraordinarily low pH (0–1) from one portal and substantial fluxes of Zn and Cu from waste piles in several other locations. The flux-reservoir diagram for Iron Mountain is shown in Figure 3.6.

Many other sources of metals can occur in addition to effluent and leachate waters. Metal contamination can come from smelter emissions, eroded tailings and waste rock, and transport of contaminated soils and sediment from the mining and mineral processing sites. The knowledge

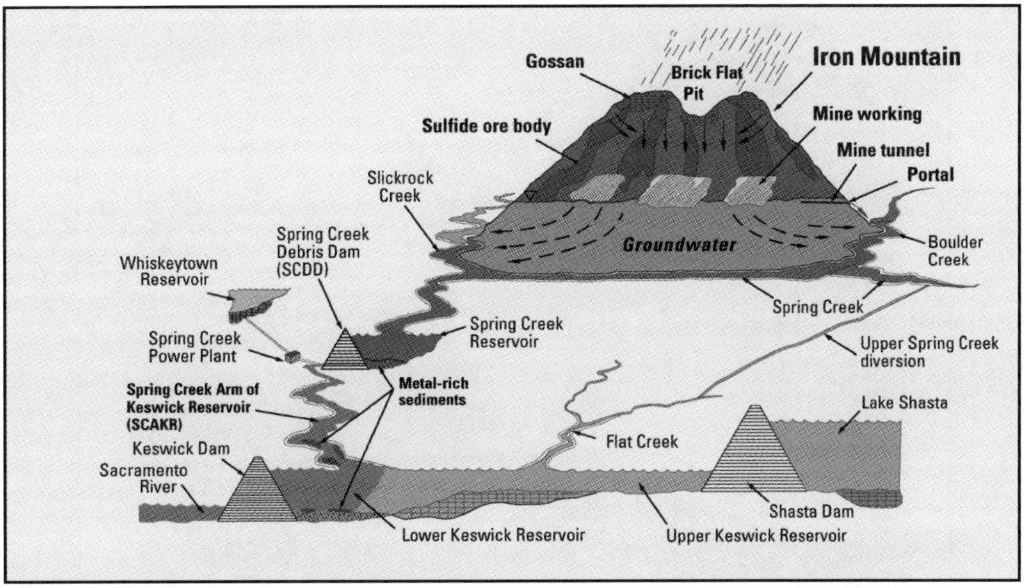

Courtesy of U.S. Environmental Protection Agency

FIGURE 3.5 Schematic diagram showing the main sources of metal contaminants from massive sulfide deposits mined underground discharging from mine tunnel portals into Boulder Creek, a tributary of Spring Creek, which is a tributary of the Sacramento River. Some metals are also derived from groundwater and leached from waste rock into Slickrock Creek.

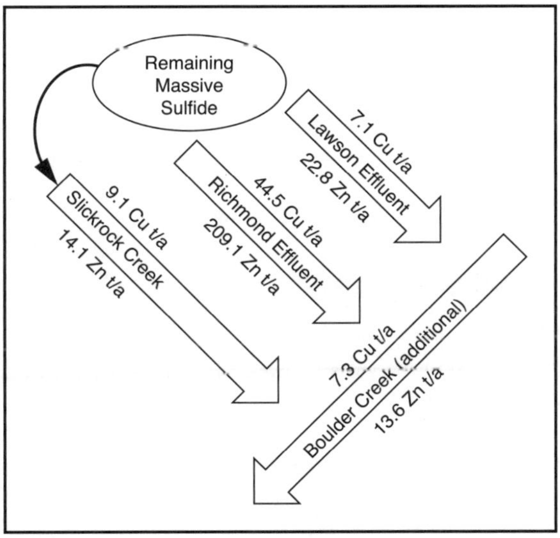

Courtesy of U.S. Geological Survey, based on data from Alpers et al. 2003

FIGURE 3.6 Simplified flux-reservoir diagram for the main Cu and Zn fluxes from Iron Mountain

of multiple sources and multiple processes to transport these sources constitutes the conceptual model and can be a leading cause of the relative success or failure of environmentally responsible mine site management. Moore and Luoma (1990) provide a generalized diagram of many of these sources and processes (Figure 3.7).

Primary	Secondary
A1. Waste Rock	B1. Groundwater at Open Pits
A2. Tailings	B2. Groundwater Beneath Ponds
A3. Slag	B3. Sediment in River Channels
A4. Underground Workings	B4. Floodplain Sediment/Soil
	B5. Reservoir Sediment

Tertiary
C1. River Sediment Reworked from Floodplain
C2. Groundwater from Contaminated Reservoir Sediment
C3. Soils from Air Pollution

Adapted from Moore and Luoma 1990

FIGURE 3.7 Examples of multiple sources and transport processes that lead to metal contaminant mobility from a mine site

Speciation Mass Balances

When describing aqueous chemical reactions, quantitative calculations are only possible if the distribution of aqueous species among free ions, ion pairs, and other complexes is known. The analytical determination of any given dissolved constituent is the sum of its species and must also obey the law of mass conservation. Consider a NaCl solution containing gypsum crystals at solubility equilibrium. A chemical analysis of the solution would be given in terms of the concentrations of Na, Ca, Cl, and SO_4. The aqueous species are Na^+, Cl^-, Ca^{2+}, $CaCl^+$, SO_4^{2-}, $CaSO_4^0$, and $NaSO_4^-$. Mass balance equations would be as follows:

$$m_{Na} = m_{Na^+} + m_{NaSO_4^-}$$

$$m_{Ca} = m_{Ca^{2+}} + m_{CaSO_4^0} + m_{CaCl^+}$$

$$m_{SO_4} = m_{SO_4^{2-}} + m_{CaSO_4^0}$$

$$m_{Cl} = m_{Cl^-} + m_{CaCl^+}$$

where m refers to molal concentrations. To determine the distribution of aqueous species under steady-state conditions, a calculation would have to be made assuming the solution was at equilibrium. The calculation uses equilibrium constants (or free energies) based on the law of mass action and activity coefficients. Most geochemical codes routinely calculate aqueous speciation because it is fundamental to more complex problems involving mineral precipitation and dissolution.

Examples of speciation calculations are provided for four waters of acid mine drainage (AMD) with compositions shown in Table 3.1.

The speciation results for copper, sulfate, aluminum, and iron are shown in Figure 3.8. The results are arranged so that the effect of different pH values and sulfate concentrations is readily apparent. As pH decreases and sulfate concentration increases, the dominant copper species

TABLE 3.1 Water analyses for four acid mine waters ranging in pH from 4.9 to 0.48

	AMD-A	AMD-B	AMD-C	AMD-D
Sample number	82WA109	82WA110	76WA103	90WA103
Water temperature	16.0°C (60.8°F)	19.5°C (67.1°F)	24.0°C (75.2°F)	34.8°C (94.6°F)
pH (field)	4.9	3.25	1.10	0.48
Constituent, mg/L				
Ca	44.7	82.2	240	183
Mg	13.5	23.6	720	821
Na	8.6	11.8	79.0	251
K	3.21	4.57	107	261
Cl	1.0	1.1	2.0	N/D*
SO_4	206	483	50,000	118,000
SiO_2	42.6	46.4	140	165
Ba	0.042	0.048	N/D	0.068
Al	5.06	19.8	1,410	2,210
F	0.30	0.52	2.0	N/D
Fe(T)	4.72	18.4	11,000	20,300
Fe(II)	4.44	9.01	7,820	18,100
Fe(III)	0.28	9.39	3,180	2,200
Mn	1.26	3.04	11.0	17.1
Cu	0.09	0.23	360	290
Zn	0.04	0.15	1,860	2,010
Cd	0.004	0.01	14.0	15.9
As(T)	0.01	0.02	N/D	56.4
As(III)	N/D	N/D	N/D	8.14
As(V)	N/D	N/D	N/D	48.3

Adapted from Alpers and Nordstrom 1999
*N/D = not determined

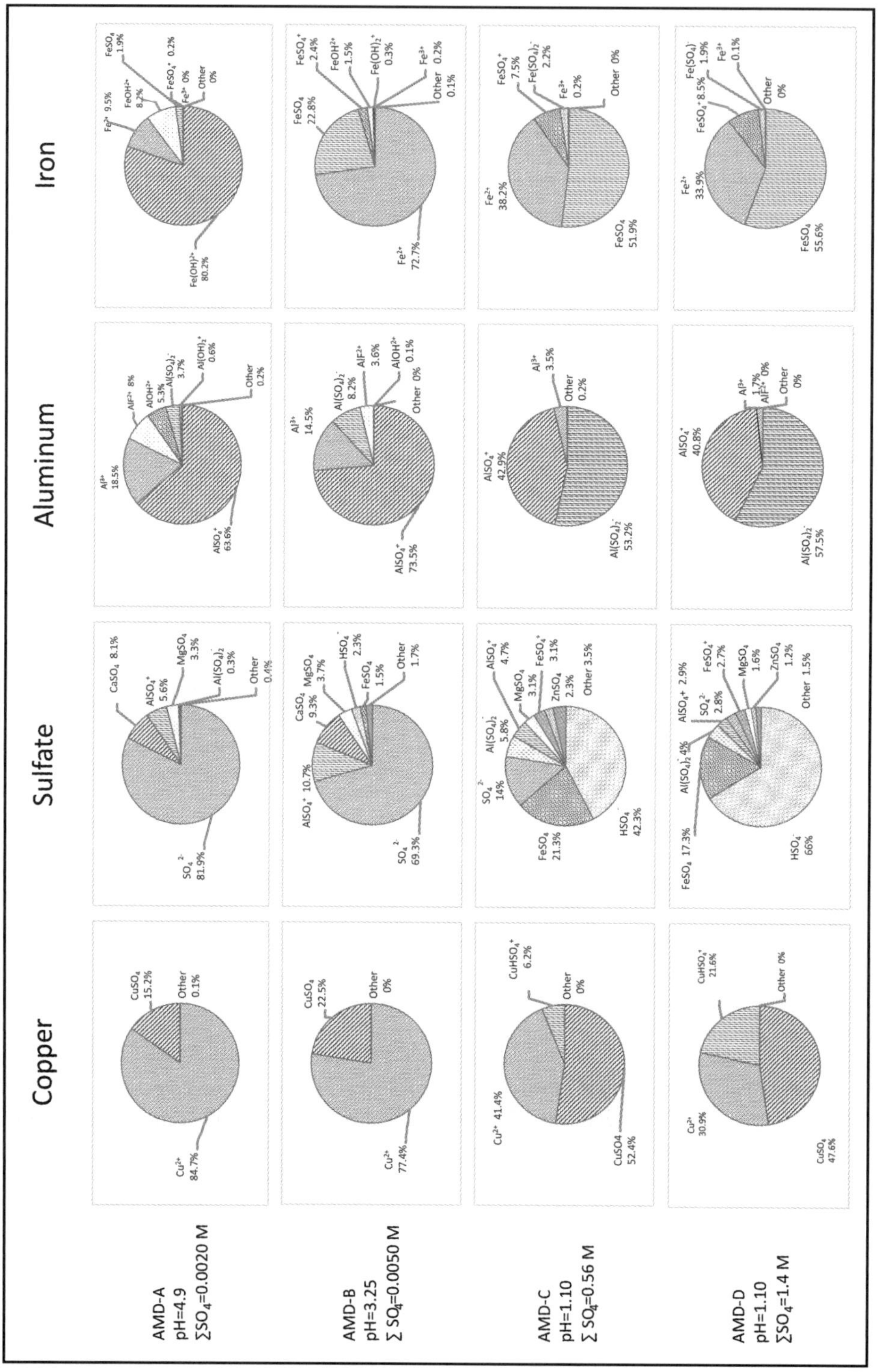

Data computed with The Geochemist's Workbench software.

FIGURE 3.8 Pie charts showing results of aqueous speciation calculations in terms of relative molalities of copper, sulfate, aluminum, and iron species using The Geochemist's Workbench with the PHREEQC database for acid mine water compositions in Table 3.1 (adapted from Alpers and Nordstrom 1999)

changes from the free cupric ion to the $CuSO_4^0$ ion pair. For sulfate, hydrolysis occurs, promoting the reaction: $SO_4^{2-} + H^+ \rightarrow HSO_4^-$ with decreasing pH. The ion pairing of sulfate for aluminum is quite strong so that the $AlSO_4^+$ complex is dominant over the whole range of pH and compositions. The speciation for iron is complex because it not only depends on the oxidation state of the dissolved iron, but also on the concentration of sulfate and other metal cations.

Activity Coefficients

The evaluation of the reactivity of aqueous solutes might be much simpler if these calculations only involved the determination of species distributed among free ions, ion pairs, triplets, and so forth. Such complexes are formed by simple electrostatic or attractive forces between ions of opposite charge. However, there are other forces at play in electrolyte solutions. These forces include interactions between the solute and the solvent (water) and repulsive forces between like-charged ions. All of these interactions lead to nonideal behavior. For example, if electrolytes behaved ideally, the solubility of a mineral like barite, $BaSO_4$, could be simply described in terms of the dissolved Ba and SO_4 concentrations, $m_{Ba^{2+}}$ and $m_{SO_4^{2-}}$. However, because of nonideal behavior, the effective or reactive concentration is something less than the actual concentration determined analytically. The effective concentration is known as the activity, a_i, for solute or component, i. The deviation of the activity from the measured molal concentration, m_i, is called the activity coefficient, γ_i:

$$\gamma_i = \frac{a_i}{m_i} \qquad \text{(EQ 3.13)}$$

There are several equations that estimate the activity coefficient based on the ionic strength, I, of the solution. The ionic strength is half the sum of all the concentrations of ions in solution multiplied by their charge squared:

$$I = \frac{1}{2}\sum_i m_i z_i^2 \qquad \text{(EQ 3.14)}$$

Activity coefficient equations have limitations in the range of concentration and composition of solution that they can be used for. There are two general types: extended Debye–Hückel and Pitzer equations. The Pitzer equations allow activity coefficient calculations to quite high ionic strengths but are limited in the range of solutes for which they can be applied. The extended Debye–Hückel equations can be used for nearly any solute but are usually limited to about 1 molal in ionic strength, which is useful for most water samples except brines. Several computer codes are available that utilize both the extended Debye–Hückel or ion-association model and the Pitzer model. These codes compute both the speciation and activity corrections for solutes found in natural waters. More details can be found in reviews and summaries of these methods of calculation (Nordstrom and Campbell 2014; Nordstrom and Munoz 1994; Millero 2001; Langmuir 1997).

Saturation Indices

Once ion activities have been calculated, it is a simple matter to calculate the saturation index. This index describes the thermodynamic tendency of a water to dissolve or precipitate minerals. If calcite dissolves in water, the reaction is

$$CaCO_3 \rightarrow Ca^{2+}_{(aq)} + CO^{2-}_{3(aq)} \qquad \text{(EQ 3.15)}$$

and when solubility equilibrium has been reached, the equilibrium constant expression is

$$K_{sp} = \frac{\left(a_{\mathrm{Ca2+}}\right)\left(a_{\mathrm{(CO3)2-}}\right)}{a_{\mathrm{CaCO3}}} \quad \text{(EQ 3.16)}$$

If the calcite is pure and well crystallized, its activity is unity and the equilibrium constant expression becomes a simple ion-activity product, IAP, which is the K_{sp} or solubility product constant. An IAP can be calculated for any water composition if all the major ions have been determined and the analytical results and the thermodynamic data are of good quality. To determine if the solution composition has reached saturation with respect to any given mineral, the IAP is compared to the K_{sp} for equilibrium solubility:

$$\frac{\mathrm{IAP}}{K_{sp}} \quad \text{(EQ 3.17)}$$

which would equal 1 at equilibrium. The saturation index (SI) is the logarithm of this quantity:

$$\mathrm{SI} = \log\frac{\mathrm{IAP}}{K_{sp}} \quad \text{(EQ 3.18)}$$

such that at equilibrium, SI = 0; for supersaturation, SI > 0; and for undersaturation, SI < 0. If supersaturation is reached, the water can precipitate the respective mineral. Whether it actually does or not depends on precipitation kinetics or the rate at which the mineral would precipitate under the specific conditions. Likewise, if a mineral is present but the SI indicates undersaturation, the mineral can dissolve, but whether it does or not depends on the dissolution kinetics. If SIs are combined with inverse modeling, a much more quantitative interpretation is possible for water–rock interactions in both surface and groundwaters.

Water-Mineral Mass Balances (Inverse Modeling)

As water flows downgradient in an aquifer, the composition of the water may change depending on the recharge conditions, mineral reactions occurring in the aquifer, and the redox (or oxidation–reduction) conditions of the aquifer. Constituents are partitioned between the aqueous phase and the solid phases in an attempt to readjust to changing chemical conditions. Minerals may dissolve or precipitate; gases may form or be consumed; organic matter may be degraded; and dissolved constituents may sorb, precipitate, or be degraded. The mass balance method that accounts for these changes conceptually consists of the following net chemical reaction:

$$\begin{array}{c}\text{initial solution composition} + \text{reactant phases} \rightarrow \\ \text{final solution composition} + \text{product phases}\end{array} \quad \text{(EQ 3.19)}$$

This reaction can be represented mathematically by the following equation:

$$\left\{\sum_{p=1}^{P} \alpha_p \beta_{p,k} = \Delta m_{\mathrm{T},k}\right\}_{k=1,j} \quad \text{(EQ 3.20)}$$

TABLE 3.2 Concentrations of groundwater from a limestone aquifer

Elements in Groundwater	Final (Groundwater), mg/L	Initial (Rainwater), mg/L
Dissolved inorganic carbon	237	0.9
Ca	50	0.001
Mg	25	0.001

Adapted from Langmuir 1971

in which α_p is the net mass transfer in moles of the *p*th plausible phase among the *P* total reactant and product phases in the reactions; $\beta_{p,k}$ is the stoichiometric coefficient of the *k*th element in the *p*th phase; Δ denotes the difference between the final and initial values; and $m_{T,k}$ is the total molality of the *k*th element in solution (Parkhurst and Plummer 1993; Plummer and Back 1980; Plummer et al. 1983). These mass balances can be solved by a simultaneous solution of a set of equations in which the number of aqueous solutes equals the number of phases, as shown by the following matrix formulation:

$$\begin{bmatrix} \beta_{1,1} & \beta_{1,2} & \beta_{1,3} & \cdots & \beta_{1,P} \\ \beta_{2,1} & \beta_{2,2} & \beta_{2,3} & \cdots & \beta_{2,P} \\ \beta_{3,1} & \beta_{3,2} & \beta_{3,3} & \cdots & \beta_{3,P} \\ \cdot & \cdot & \cdot & \cdots & \cdot \\ \cdot & \cdot & \cdot & \cdots & \cdot \\ \cdot & \cdot & \cdot & \cdots & \cdot \\ \beta_{j,1} & \beta_{j,2} & \beta_{j,3} & \cdots & \beta_{j,P} \end{bmatrix} \begin{bmatrix} \alpha_1 \\ \alpha_1 \\ \alpha_1 \\ \cdot \\ \cdot \\ \cdot \\ \alpha_P \end{bmatrix} \begin{bmatrix} \Delta m_1 \\ \Delta m_1 \\ \Delta m_1 \\ \cdot \\ \cdot \\ \cdot \\ \Delta m_j \end{bmatrix} \quad \text{(EQ 3.21)}$$

where *P* is the total number of plausible phases and *j* is the total number of elements or components. A simple example will clarify this method.

Consider a limestone aquifer in a recharge area. We can assume infiltrating rainwater to be at equilibrium with atmospheric CO_2 ($\log P_{CO_2} = -3.5$). That equilibrium condition gives a DIC (dissolved inorganic carbon) concentration in rainwater of 0.0107 mmol $kg^{-1}_{H_2O}$. According to this mass balance (partitioning of total C between gas and liquid):

$$CO_{2(g)} + H_2O \rightarrow H_2CO_{3(aq)} \quad \text{(EQ 3.22)}$$

Next, rainwater infiltrates and dissolves some of the rock to produce the concentrations presented in Table 3.2. These are actual concentrations (in milligrams per liter) of a groundwater from a well in a limestone aquifer (Langmuir 1971).

The only reacting minerals that need to be considered in the aquifer are calcite and dolomite. Calcium can dissolve from both calcite and dolomite, Mg can only dissolve from dolomite, and carbon dioxide (CO_2) comes from rainwater plus calcite and dolomite dissolution. The CO_2 in rainwater was chosen to be in equilibrium with atmospheric $CO_{2(g)}$. Simultaneously solving for the amounts of these minerals dissolved by rainwater gives the mass balance model shown in Table 3.3.

The overall reaction can then be written as

$$\text{rain} + 1.59CO_2 + 0.022CaCO_3 + 1.03CaMg(CO_3)_2 \rightarrow \text{well water composition} \quad \text{(EQ 3.23)}$$

TABLE 3.3 Mass balance model

Minerals and Gases	Mass Transfer, mmol/kg_{H_2O}
CO_2	1.59
Calcite	0.022
Dolomite	1.03

Another model is also possible with no uptake in CO_2, but it would seem unusual for infiltrating rainwater not to take up any soil CO_2.

Next we shall consider a mass balance performed on another actual water analysis in which the dominant reaction is pyrite oxidation. The water analysis is taken from the report by Ball and Nordstrom (1989) for effluent from the Leviathan mine in California, southeast of Lake Tahoe. The mineralogy at this site is not well characterized, but the description fits that of a near-surface hot spring environment with a large layer of elemental sulfur lying on top of, and adjacent to, an andesitic tuff (Evans 1966). It was primarily mined for sulfur to make sulfuric acid. Assuming pyrite, silica sinter, kaolinite, K-feldspar, andesine, and chlorite were present, and that silica and goethite can precipitate, a mass balance calculation was conducted for major ions in the effluent water of the No. 5 tunnel. It was shown by trace-element determinations and by sulfur isotopes that nearly all the sulfuric acid was produced from pyrite oxidation, not from sulfur oxidation (Ball and Nordstrom 1985). The composition of the water is shown in Table 3.4. A mass balance was calculated (Nordstrom 2011) and the results are shown in Table 3.5. It was assumed that the initial state was pure water and the final state was the effluent water in Table 3.4. Several minor and trace elements were left out of the calculation because the exact mineral source is not known. Minerals that dissolved have positive values, and minerals that precipitated have negative values. This calculation is consistent with every other calculation for acid mine water draining from a mine or waste rock in that almost everything dissolves and silica and some form of iron mineral always precipitate. At the time the water sample was collected, the discharge was about 1 L/s, which converts to about 141 moles of pyrite weathering per hour after accounting for mass balances.

The next example is natural acid rock drainage from Straight Creek, northern New Mexico, for which more detailed mineralogy was obtained (Plumlee et al. 2005). The water composition was fairly constant over nearly two years of sampling, and a median composition and average discharge were selected for the mass balance evaluation. The water composition is shown in Table 3.6 and the mass balance results in Table 3.7.

These results, when combined with a median flow of 37.85 L/min for Straight Creek drainage, give an approximate weathering rate of 20 moles FeS_2/h, which is about 3–5 orders of magnitude higher than lab studies on the microbial oxidation rates of pyrite (about 0.22–20 mmol/L/h; Nordstrom 2011). However, it is important to realize that the former is a flux rate and the latter is a reaction rate only. A flux or transport rate can include other reactions and processes in addition to the pyrite oxidation rate, such as precipitation and dissolution of soluble sulfate salts, storage of evaporating acid waters, and so on.

TABLE 3.4 Water analysis of Leviathan acid mine drainage (No. 5 tunnel)

Determination	Value
Temperature	12.0°C (53.6°F)
pH	1.80
Specific conductance	9.3 mS/cm
Concentrations, mg/L	
Aluminum	438
Cadmium	0.282
Calcium	131
Chloride	8.4
Chromium	2.58
Cobalt	5.11
Copper	5.32
Ferrous iron	1,440
Fluoride	3.4
Total iron	1,570
Lead	0.0745
Magnesium	53.8
Manganese	10.5
Molybdenum	0.024
Nickel	11.9
Silica	109
Sodium	19.4
Strontium	2.84
Sulfate	7,540
Thallium	2.0
Vanadium	1.23
Zinc	1.39

Source: Ball and Nordstrom 1989

TABLE 3.5 Mineral mass balance calculations based on the water analysis from Table 3.4

Mineral Phase	Model Result, $mmol/kg_{H_2O}$
Pyrite	39.25
Kaolinite	6.78
Calcite	2.87
K-spar	0.353
Andesine	1.005
Chlorite	0.442
NaCl	0.240
Goethite	–11.25
Silica	–14.7
pH	1.80

Source: Nordstrom 2011

TABLE 3.6 Water analysis for average (median) Straight Creek natural acid rock drainage

Determination	Median Value for *n* = 14
pH	2.98
pH Lab	2.70
Specific conductance	3,085 μS/cm
Specific conductance lab	3,200 μS/cm
Temperature	7.7°C (45.9°F)
Eh	0.781
Dissolved oxygen	5.59
Constituent, mg/L	
Ca	349
Mg	113
Na	8.08
K	0.753
SO_4	2030
F	7.71
Cl	2.98
SiO_2	74.2
Al	91.5
Fe	65.0
Fe(II)	0.475
Li	0.181
Sr	0.612
Ba	0.002
Mn	20.8
Zn	7.63
Pb	0.012
Ni	0.727
Cu	1.87
Cd	0.039
Cr	0.040
Co	0.325
Be	0.026
Dissolved organic carbon	1.44

Source: Nordstrom 2011

TABLE 3.7 Mass balance results for water analysis of Table 3.6

Mineral Phase	Model Result, mmol/kg(H_2O)
Pyrite	8.66
Dolomite	4.64
Gypsum	3.78
Kaolinite	1.40
Oligoclase	0.44
Fluorite	0.20
Sphalerite	0.11
Illite/Sericite	0.032
Chalcopyrite	0.029
Goethite	–7.40
Silica	–2.89
pH	2.98

Source: Nordstrom 2008

CHAPTER 4

Modeling for Regulatory Review

Andrew Nicholson and D. Kirk Nordstrom

INTRODUCTION

Environmental models are applied in a variety of regulatory settings for permitting, setting compliance requirements, characterizing sites, predicting future scenarios, and evaluating remediation options. For mining sites, there is a need to predict the engineering performance of a planned mine operation, the performance of an operating mine, or the performance of possible alternative remedial measures. Because many aspects of mine permitting, operations, and closure rely on a forecast of future conditions, the application of models is essential.

Regulatory policies and potential model applications at mine sites can include a broad and diverse range so that a comprehensive list of the specific regulatory applications of environmental models would be challenging to compile. Consequently, the goal of this chapter is to provide some guidance on the application of models in a regulatory setting to aid experts in technical, regulatory, legal, and economic fields so that they might work together to create more effective models at mining sites.

GOOD SCIENCE

Meeting regulatory requirements may keep mining and remedial activities legal, but such practice does not necessarily ensure the most cost-effective remediation and long-term prevention of future contamination. The path to effective remediation over both the short and long term is through proper site characterization that quantifies as much as is practical. The dominant hydrogeologic and geochemical processes should be outlined, as in a flux-reservoir diagram, followed by an evaluation of remedial scenarios using models that are based on properly collected field data and sound professional judgment. To achieve these goals, good science is essential, science that utilizes geological, hydrological, and geochemical expertise. Depending on circumstances, other expertise in biology, microbiology, hydrometallurgy, and land use management may be needed as well.

Characterization and remediation most commonly fail, and/or cost overruns occur, when the following take place:

1. Deadlines become more important than the evaluation of site characteristics and careful evaluation of remedial options.
2. Default technological approaches are embraced prior to proper system characterization.
3. No survey of the successes or failures of default approaches has been conducted.
4. The costs, results, or implications of scientific characterization are feared by parties working at sites.

5. Monitoring is not attempted or adequately achieved for both spatial and temporal variations.
6. Stakeholders involved in the planning of the characterization or remediation are not adequately versed in the issues at hand.
7. Personnel who carry out the field, analytical, and interpretive tasks are not adequately trained and experienced in proper procedures.
8. No independent technical oversight committee exists to offer critiques and suggestions during the characterization and remediation activities.

A considerable amount of research relevant to remedial needs is available for practical application, and the challenge is to get that information to remediation specialists and to regulatory agencies. Some of this information activity is available through the International Network for Acid Prevention (INAP), Mine Environment Neutral Drainage (MEND), Acid Drainage Technology Initiative–Metal Mining Sector (ADTI-MMS), and U.S. Environmental Protection Agency (USEPA), but there is still a scarcity of options related to recycling and waste as a resource. Summaries of the diversity of remedial options should be readily accessible and regularly updated. Comparisons of predicted with actual monitored consequences (such as Kuipers et al. 2006) should be available and regularly updated to make the necessary improvement in characterization, remediation, and monitoring of mine sites. Monitoring a site before, during, and after final reclamation and closure is often inadequate to even determine what difference was made by the remediation.

The goal of site characterization should be to understand the major hydrogeochemical processes causing or likely to cause migration of constituents potentially affecting environmental and human health. If this is not the goal, the chances of failure markedly increase. If it is the goal, regulatory compliance becomes easier to achieve and remediation will be more effective. Conscientious hydrogeochemists would not consider serious interpretations of these processes without the inclusion of modeling to aid in the prediction of future environmental conditions.

REGULATORY CONCERNS

Modeling in support of regulatory decision making has some specific challenges in implementation. The questions being evaluated are often not strictly scientific in scope, nor are they free from politicization. The political implications surrounding environmental models should not be ignored. In the process of permitting mines, mining companies have an interest in rapidly moving from a preproduction stage where there is no positive cash flow to a production stage where there is income from a mine. Hence, they can be motivated to model optimistic outcomes and best-case scenarios. Regulators may also have incentives to facilitate mine permitting, which can result in poorly evaluated models submitted in support of permitting. Opponents to mining may produce models that overpredict environmental impacts and present worst-case scenarios to illustrate their concerns.

The needs of regulation, industry, and science are often competing in these situations. Regulation can drive the modeling process by prescribing the codes and methods to use, which may exclude other more scientifically rigorous approaches. For example, the regulations to determine soil remediation levels protective of groundwater in some states prescribe specific codes and input assumptions for the calculation of site-specific remedial goals. Montana prescribes the application of HELP (Hydrologic Evaluation of Landfill Performance; Schroeder et al. 1994) and VS2D (Lappala et al. 1987) computer programs with some recommended input parameters to set these levels. California has developed some specific spreadsheets with assumptions regarding partitioning and infiltration rates to calculate these levels. Frequently, these prescribed assumptions are conservative in nature, designed to be protective of the environment rather than to get the best estimate of what might actually happen. Budget constraints by project proponents can limit the depth and approach of some modeling exercises, limiting the adequacy of characterization or remedial options. The legal, scientific, engineering, and commercial aims and goals may, therefore, not be compatible. Many regulatory agencies have developed guidance documents that have additional information on the application and guidance of models in regulatory settings (e.g., USEPA 2009; NRC 2007).

Modeling and Data Collection

Adequate characterization of sites is often the most important prerequisite for understanding site hydrogeochemistry. As previously discussed, models can be used to understand site data, guide the collection of further data, or suggest future behavior based on the data. When models are being used for regulatory decision making, it is essential to identify whether the data are adequate to support the modeling and evaluate how limitations on the data affect the certainty of model results.

The data needs for a model will depend on the type of model. Models need to be based on data, and the limitations of the data need to be explicitly stated and clearly evaluated in the modeling process. At their best, models allow for the extraction of the most important information from the available data. In general, the higher the quantity and quality of data available, the more reliable will be the results of the model. One classic example in mine development is the block models developed to predict ore reserves. In this case, the more extensive the data collection from surface mapping and drill cores to interpret the ore deposit, the more reliable will be the geostatistical estimates of the ore grade of the potential mine. Furthermore, these data can often provide valuable estimates of environmental concerns. However, limited data is almost always the case in environmental systems. In many modeling applications, it may not be possible for adequate data to be collected because the cost of data collection and the variability of the measured values are too great. Some models are adequately supported with very little data. For example, a handful of experiments may provide sufficient data to effectively model a water treatment process. Other models will always have data limitations because the variability of field conditions is so great, as is the case with coupled groundwater-chemistry flow models. In these situations, evaluating the uncertainty in the model due to data uncertainty is important and sensitivity analysis is usually necessary.

With modern geographic information systems, data analysis, and modeling techniques, models themselves and supporting information can be used to better characterize data gaps and model uncertainties to optimize data collection techniques. These include

- Three-dimensional mapping and analysis of available data;
- Application of probabilistic modeling approaches to evaluate data uncertainty; and
- Use of codes such as PEST (Doherty 2010; Doherty and Hunt 2010; Dausman et al. 2010) to evaluate model uncertainty, identify the most sensitive parameters, and optimize the areas of additional data collection.

Model Formulation

Prior to embarking on a modeling effort, the expectations of the modeling effort should be clear, either through established regulation and precedent or by means of dialog with stakeholders.

Environmental models often support regulatory decisions, and the model application should be developed with an eye toward effectively answering the regulatory issues being addressed. For example, it may make sense to have a higher resolution modeling effort (and collect more data) near sensitive features that a model is addressing. If a model is being used to assess future conditions (which is almost always the case), the appropriate time frames for modeling need to be established and the model may need to be refined with the incorporation of additional data as a mine operates. For periods of time well beyond a calibration or history-matching period, modeling will be highly uncertain unless well-documented analogue conditions have undergone prediction followed by monitoring for many years, which would demonstrate that predictions indeed fit long-term observational data. For these long-term predictions, modeling is so uncertain as to be an unreliable approach to decision making. Alternative approaches must be evaluated that do not depend on modeling. Expectations for what needs to be included in the model must be established and a process for evaluating and updating the model included as part of establishing a pathway forward on the project. One example is the common Nevada Division of Environmental Protection requirement that pit lake models be updated every five years at mines (http://ndep.nv.gov/bmrr/regs.htm) as more data are collected and the deposits are better understood with ongoing mining.

A parameterized model is formulated based on a conceptual model for the site. The conceptual model is a dynamic conceptualization that changes with more quantity and quality of data and with expert discussion over the interpretation of the data. The importance of the conceptual model cannot be overemphasized because it guides the decision-making process and the data collection.

Code Selection

The selection of a modeling code is essential in the regulatory application of models. The code being used should be

- Based on accepted principles and established methods,
- Available and provided with the model, and
- Tested using standard cases and compared to the results of other codes for a similar range of conditions applicable to the mine site.

Use of proprietary codes is sometimes necessary, but such a code must be provided to reviewers to ensure that the model is behaving as represented, the database is appropriate, and the necessary assumptions are understood.

Model Testing

Testing of models and model results should include the following:

- Comparison against the conceptual model for the site
- Comparison against simplified assumptions
- Vetting of mass balances to ensure they make sense
- Acceptable replication of model scenarios
- Consideration of extreme conditions applicable to the site
- Evaluation of both best-case and worst-case possibilities for future scenarios

Effective Presentation

Models and associated codes need to be presented to stakeholders in an understandable, transparent fashion. Important aspects include

- Written documentation of the model development, underlying assumptions, implementation, and application;

- The code being used either publically available or provided to reviewers in a useful and transparent form;
- The provision of all relevant input files and underlying data to reviewers and stakeholders; and
- Explicit recognition of the uncertainties and limitations of the model results.

Effective presentation of models flows from good modeling practices and reflects the reasoning of the modeler and the modeler's ability to understand the data. Good presentations typically include flowcharts that describe the modeling procedure and approach; easy-to-read charts and figures that present modeling inputs and results; examples to clarify difficult concepts; and a short, concise summary of the most important modeling assumptions and findings. A well-constructed model will more easily lend itself to clear presentation and documentation than if the modeling approach has technical limits or skips steps.

Uncertainty Analysis

There are at least three main sources of uncertainty:

1. Data uncertainty
2. Assumption uncertainty
3. Uncertainty regarding whether the model and associated code have demonstrated reliability for the range of conditions presented

Uncertainties arise from a variety of sources, some of which are quantifiable and some of which are not. Potentially large uncertainties should be explicitly stated and quantified where possible. For example, data errors include precision and accuracy errors from analytical measurements of elemental composition of materials, and errors arising from sampling issues (e.g., errors in filtration, errors in misuse or malfunction of field equipment, errors in preservation). Analytical errors in accuracy and precision can be quantified if the laboratory has conducted proper quality assurance/quality control procedures. Sampling and preservation errors are more difficult to quantify or even identify. Whether enough (or too much) analytical data have been collected is another source of uncertainty, not related to laboratory capability but to the project objectives, scope of work, and general planning. Propagation of analytical errors through model calculations provides some concept of the effect of analytical uncertainties on the study results.

Thermodynamic and kinetic databases can be sources of uncertainty for geochemical modeling. To avoid any ambiguities, it is easiest to compile and report the thermodynamic and kinetic data used in the modeling. At the very least, references should be provided that allow easy access to the sources of the relevant thermodynamic and kinetic data.

When modeling, assumptions can constitute the largest source of uncertainties. Modeling assumptions must be clearly and explicitly stated. When quantifiable by some bounding conditions, these limits should be stated.

Model Review

Who is the reviewer? The reviewer of any model obviously needs to have the skills and experience to appropriately understand the workings of the model. Although this statement may seem obvious, putting it into practice is often challenging.

The technical capability of the reviewer is the first qualification for model review. The reviewer should know the underlying technical principles of the model and be experienced with the type of modeling being done. He or she should have a combination of education, experience, and documented peer-reviewed publications demonstrating his or her expertise in the area being reviewed. The technical expertise of the reviewer should be at a level similar to the experience and expertise of that required to produce the model.

It may take more than one reviewer to appropriately review a model, especially with models that cross disciplines. For example, a model of flow and transport of acid rock drainage in a waste rock dump that is being capped may be most appropriately reviewed by a combination of an engineer familiar with the construction of caps, a geochemist who understands the chemistry of acid generation and contaminant mobility, and a hydrologist who understands flow in the unsaturated zone. These experts should have documented experience in their respective areas of expertise and ideally would have experience with the specific code being used. If a proprietary code is being used or a poorly documented public code, it is critical that the reviewers have the expertise to review the function of the underlying code itself, beyond reviewing input and output files. It may be necessary to run the input data on another publicly available, peer-reviewed code to determine the adequacy and reliability of the modeling.

The reviewers should not have a conflict of interest in the project being reviewed. With the consolidation of mining companies and the fact that the environmental mining sector consists of a relatively small group of practitioners, this can be a challenge, as many of the most qualified scientists and engineers available for review will have worked with many of the companies developing mines at some point in their careers. Acknowledging this fact, it is important that reviewers not have a financial interest in the project and disclose their previous working relationships related to the mining industry, including work for mining companies, regulatory agencies, nongovernmental organizations, and any other third parties active in the area of the social/environmental aspects of mining.

Adequate model review requires that the reviewers are appropriately compensated for their work, recognizing the level of effort required for the review. With the importance of the underlying conceptual models and the specifics driven by site conditions, it is essential that reviewers visit the site to provide a better understanding of the local conditions that may affect model results. The reviewers must also recognize the importance of their role and conduct themselves appropriately. They should not divulge proprietary data, and the nature of what is proprietary and what is publically available data should be determined prior to the start of the model review. Reviewers also need to comply with site-specific safety requirements on site visits and should be provided with appropriate training as needed prior to site visits.

Aspects of model review. The mechanics of a model review should follow common-sense guidelines. The full code used for the modeling must be available to the reviewers, and all input files must be provided. The input files should include the full data set obtained for the site. Models must have adequate documentation both for the code and application of the code. A presentation of the modeling results for the reviewers with adequate time for questions and answers is necessary to facilitate a better understanding for the reviewers. An important rule of thumb is that if a modeler cannot explain his or her results and how they obtained the results to a nontechnical person, then the modeling analysis probably needs to be redone. In the executive summary of USEPA's *Guidance on the Development, Evaluation, and Application of Environmental Models* (USEPA 2009), the statement on model application is worth repeating here:

> *Model application (i.e., model-based decision making) is strengthened when the science underlying the model is transparent. The elements of transparency emphasized in this guidance are (a) comprehensive documentation of all aspects of a modeling project (suggested as a list of elements relevant to any modeling project) and (b) effective communication between modelers, analysts, and decision makers. This approach ensures that there is a clear rationale for using a model for a specific regulatory application.*

For regulatory decision making, review of documentation and results alone is not sufficient. Review should include comparing the model results with known test cases and actually running the software for problems of the type for which the model might be used. For example, there are many examples given in code manuals such as PHREEQC (Parkhurst and Appelo 2013) and TOUGHREACT (Xu et al. 2004) as well as documented field applications in the scientific literature that could be duplicated by the particular code being applied to show that it gives the same results. Alternative software and other models familiar to the reviewers should be used to give the reviewers a second opinion and the confidence that the results are at least consistent with other approaches.

When models and codes are used to calculate predictions for tens to hundreds or even thousands of years into the future, it must be understood that the processes underlying the model are likely to be no longer accurate or even meaningful due to changes in climate, erosion/sedimentation, anthropogenic activities, subsidence, volcanic activity, or other factors that are not identified. Such long-term future predictions are largely guesses and should be carefully qualified, as Pilkey and Pilkey-Jarvis (2007) have pointed out. The reason these long-term models have limited value is that, due to our inability to test the predictions after decades or centuries have passed, we do not know if the prediction is reliable or not, which is generally outside good scientific and engineering practice. Therefore, such predictions are limited and sometimes untrustworthy in their ability to help decide important policies regarding the future of our soil, air, or water resources. They should be designated "untestable computational experiments" or "future scenario guesses" until such time that we have good monitoring data over several decades that can be compared to the original predictions. Model computations can sometimes put bounds on the possible range of future conditions of a site, but there are always large uncertainties, for instance, because of our inability to predict climate change quantitatively for any given location and our inability to predict changes in population density and water usage far into the future. There are alternatives to model scenario predictions on which to base regulatory policy, which should be utilized more often. Alternatives may include natural analogues, consensus of expert opinion, and intentional temporary solutions awaiting further technological refinement. The question of uncertainties and the propagation of errors from these uncertainties is a challenging subject when it comes to complex environmental sites. It would seem to be fundamental to any future scenario speculation, but it is not routinely done. It is possible to consider what input variables a model computation is most sensitive to and consider a range of uncertainty or error on each of them, and then bias them for the best-case and worst-case scenarios to determine if the model calculation offers sufficient constraints to be useful or not. Then the peer review committee can evaluate that range of scenarios.

Again, quoting from the USEPA (2009) guidance on modeling:

> *Model evaluation is the process for generating information over the life cycle of the project that helps determine whether a model and its analytical results are of sufficient quality to serve as the basis for a decision. Model quality is an attribute that is meaningful only within the context of a specific model application. In simple terms, model evaluation provides information to help answer the following questions: (a) How have the principles of sound science been addressed during model development? (b) How is the choice of model supported by the quantity and quality of available data? (c) How closely does the model approximate the real system of interest? (d) How well does the model perform the specified task while meeting the objectives set by quality assurance project planning?*

CONCLUSIONS

Geochemical modeling is a complex process when applied to mine sites to discern, explain, or predict water chemistry changes over time and space. It can be applied to site characterization and to potential alternatives to remediation when simulation scenarios are examined. That is, different remedial scenarios can be considered and the water quality consequences estimated along with a cost–benefit analysis of each (such as the Summitville mine restoration planning, Ball et al. 2004). Key parts of the model development are (1) matching the model with the objectives, (2) capturing and understanding the important conditions and parameters in the model, (3) developing a database that meets quality objectives for the model and general quality standards, (4) testing the model for application similar to objectives for which it is intended, and (5) predicting conditions that can be tested with additional independent observations. Model application requires appropriate expertise on the part of the modeler, sufficient data of the right type, sensitivity analysis, uncertainty analysis, and, for future predictions, consideration of an adequate range of conditions that have undergone robust sensitivity and uncertainty analysis. Model evaluation requires a peer review team void of obvious conflicts of interest. It requires transparency of any codes, adequate documentation for codes and models, availability of all input data, and effective communication among all stakeholders.

CHAPTER 5

Modeling to Understand a Site

D. Kirk Nordstrom

INTRODUCTION

Modeling begins with a conceptual framework for a site and specific objectives that can be converted into tasks within a scope of work. To complete these tasks, field data and samples must be collected and analyzed by others and will be relied upon, sometimes heavily, by modelers. Hence, there should be a close collaboration between those collecting, analyzing, and vetting the data with those modeling, and this collaborative process should start at the beginning of a project. The following section begins with an introduction to the importance of obtaining reliable water analyses before proceeding to how modeling can interpret water chemistry at a site.

WATER ANALYSES

Geochemical modeling is best constrained and most useful when actual water and mineral analyses are used. However, sources of error and uncertainty can occur anywhere—from the time, place, and protocols of sampling to the instrumental technique and the quality assurance/quality control (QA/QC) procedures at the analytical laboratory (McLemore et al. 2014). Because any modeling interpretations are highly dependent on the quality of the data, a summary review of sampling and QA/QC procedures is warranted here. More details can be found in Volume 6 of this handbook series.

Water Sample Collection

"Sampling is a vital part of studies of natural-water composition and is perhaps the major source of error in the whole process of obtaining water-quality information" (Hem 1992). Water chemistry is affected by several factors that are not always considered in the planning of a water sampling campaign. For surface waters, it can change with width and depth of a stream (or lake) as well as with distance from source and with time of day, day of year, and year. If acid rock drainage (ARD) discharges into a larger stream or river, it tends to hug that side of the river where it is entering and it could be transported meters to many kilometers downstream before it is fully mixed. The classic example is the Susquehanna River, which has been monitored at Harrisburg, Pennsylvania (e.g., P.W. Anderson 1963; Fishel 1984), and where substantial chemical gradients were noted across the width of both east and west channels. Sulfate concentrations were higher toward each bank because of the ARD entering the river for kilometers upstream. The same effect has been observed in small streams and rivers as well. With such a strong chemical gradient across the width of the river, sampling must be done with integrating composite samplers to get the average composition at any one location. Sampling at regular intervals during a 24-hour period (diel variability) will show further changes for a constant flow. Water composition also changes with flow, with dependence on whether it is increasing flow or decreasing flow; with dependence on whether it is early, middle, or late increasing or decreasing flow; with the intensity and location of the rainstorm or snowmelt; and with length of the preceding dry period. Again, the Susquehanna River was found to have substantial diel variations (Fishel 1984). Diel metal cycling in small streams affected by mine drainage has been reported in a series of papers (see review by Nimick et al. 2011). Chemical variability in pit lakes, natural lakes, and human-made reservoirs receiving discharges of ARD can also be dependent on timing and location. Considerable detail on pit lake and reservoir water quality and dynamics are presented in Castendyk and Eary (2009). Procedures for sampling pit lakes have been described by Gammons (2009).

Field measurements, collection, and preservation procedures and chemical analysis methods for water samples can be found on the U.S. Environmental Protection Agency's (USEPA's) website for Clean Water Act Analytical Methods (https://www.epa.gov/cwa-methods), in the U.S. Geological Survey (USGS) *National Field Manual for the Collection of Water-Quality Data* (USGS 2005–2015), and in *Standard Methods for the Examination of Water and Wastewater* (APHA et al. 2012) along with analytical methods.

Quality Assurance/Quality Control

Although numerous and reliable analytical techniques exist to determine the chemical composition of almost any type of water sample, it does not mean that potential issues and concerns are absent. Indeed, depending on the instrumental technique used, there can be serious interferences and systematic biases in analytical results. Protocols for QA/QC are spelled out in considerable detail by the USEPA, USGS (Fishman and Friedman 1989), and in several textbooks and journal articles. Basic QA/QC should include charge-balance calculations, blanks (should include field and lab blanks), spiked recoveries, and standard reference samples (some should be site specific). McCleskey et al. (2004) provide an example of QA/QC procedures and results for acid-sulfate waters and high-calcium, high-sulfate waters of neutral pH in a drainage basin containing pyrite weathering primarily from natural occurrences. A recent technique involves calculation of the electrical conductivity and comparing it to the measured conductivity as a further constraint on QA/QC (McCleskey et al. 2012; McCleskey 2013). In addition, other field and hydrogeologic considerations can be helpful (e.g., Nordstrom et al. 2010; Nordstrom and McCleskey 2012). For instance, acid-sulfate waters often contain sulfate as the dominant anion, and sulfate will correlate closely with conductivity. Also, acid mine waters usually demonstrate excellent correspondence between calculated and measured redox potential (Nordstrom 2011). Replicate samples from the same groundwater well should reflect similar chemistry from time to time with changes that should be explainable by seasonal effects or anthropogenic disturbances.

Charge-Balance Problems for Low-pH Waters

The charge-balance calculation is one of the first and simplest of QA/QC procedures to check the accuracy of a water analysis. It is based on the principle that an aqueous solution containing electrolytes, or charged ions, must be electrically neutral overall. The equivalent concentration of cations must equal that of anions:

$$\sum_i c_i = 0 = c_i^+ - c_i^- \qquad \text{(EQ 5.1)}$$

where c_i is the equivalent concentration (usually in milliequivalents per liter) for the ith ion.

The charge-balance calculation is normally obtained as a percentage by taking the difference of the equivalent concentration of cations and anions and dividing by the sum of the two times 100 (or it is divided by the average; Ball and Nordstrom 1991):

$$\left(\frac{\sum_i^n c_i^+ - \sum_i^n c_i^-}{\sum_i^n c_i^+ + \sum_i^n c_i^-}\right)\times 100 \qquad \text{(EQ 5.2)}$$

Two important potential problems arise for waters of low pH. One concern has to do with including the hydrogen-ion concentration and the other has to do with speciation.

Charge balances are routinely calculated by analytical laboratories using the major ions: Na, K, Ca, Mg, Cl, SO_4, HCO_3 ($\pm NO_3$, F). Once the pH drops below 3, the hydrogen ion becomes a major cation (about 1 milliequivalent); and for acid mine waters, Fe and Al become major cations as well. If the hydrogen-ion concentration is absent from the charge-balance calculation, the charge balance will become progressively worse as the pH decreases.

The second problem arises because aqueous complexes become dominant relative to free ions in acid-sulfate waters or in any water of higher total dissolved solids. Charge balances are calculated assuming that the analytical determinations of major ions represent the aqueous free ions in solution; for instance, that the analytically determined dissolved Al is all present as Al^{3+}. In ARD, sulfate is the dominant anion. Sulfate forms complexes, or ion pairs, with cations, and the higher the oxidation state on the cation, the stronger the complex. Hence, most of the dissolved Al in ARD is complexed with sulfate. Such reactions as

$$Al^{3+} + SO_4^{2-} \rightarrow AlSO_4^+ \qquad \text{(EQ 5.3)}$$

have the consequence of lowering the charge on the dissolved aluminum from 3+ to 1+. In addition, a certain amount of sulfate becomes incorporated into the cation sum. The overall effect of these speciation reactions is to lower the total charge of dissolved ions in solution, changing the "analytical" charge balance substantially with increasing concentration of dissolved ions, especially if the major dissolved ions are polyvalent, and shift some of the sulfate to the cations.

What about the hydrogen ion? Hydrogen ions and a few other cations (e.g., Na, K) can get shifted to the anion sum. Consider this reaction:

$$H^+ + SO_4^{2-} \rightarrow HSO_4^- \qquad \text{(EQ 5.4)}$$

in which the cation, H^+, is incorporated into the anionic species, HSO_4^-. The pK_2 for this reaction is 1.99, which means that as the pH approaches 2, increasing amounts of dissolved sulfate become bisulfate. Below pH 2, free sulfate ion is no longer dominant, but the bisulfate ion is. Also, this hydrogen is not part of the hydrogen measured by the pH electrode; it is additional hydrogen that must be added to the sulfate to produce the bisulfate ion.

These two problems mean that equivalent hydrogen-ion concentrations must be extracted from the pH measurement (a measurement of hydrogen ion activity, not concentration) for the charge balance to be applied to acid waters, and the water must be speciated before a charge balance can be calculated. Hence, a proper charge balance must be based on the results of speciation modeling using thermodynamic data. Some codes are programmed to do this (PHREEQC [Parkhurst and Appelo 1999, 2013], WATEQ4F [Ball and Nordstrom 1991], MIN3P [Mayer et al. 2012], CrunchFlow [Steefel 2007], and PFLOTRAN [Hammond et al. 2012]), but the analytical results from the laboratory do not usually get screened for charge balance unless the modeler does it and reports it. If a different code is used to calculate charge balance, assurance is needed that the calculation is performed correctly for acid waters.

The plots in Figure 5.1 show the divergence of the analytical charge imbalance from the speciated charge imbalance for 1,300 water samples affected by mine drainage. In Figure 5.1a, the hydrogen ion is absent altogether and the charge imbalances are close to zero until the pH is ≤4, and then they continuously diverge at lower pH. In Figure 5.1b, the hydrogen ion is included in the charge balance but not the speciation. The overall charge balances have improved, but the divergence still occurs for pH ≤ 2. In Figure 5.1c, the speciation has been included and now the charge imbalances are all distributed close to zero.

One other potential issue is noteworthy here. Measurements of pH as defined by the National Institute of Standards and Technology (NIST, formerly NBS, National Bureau of Standards) are only valid for measurements to about pH 1. Below that pH value, the conventions for defining pH no longer apply. Fortunately, another option is available based on the Pitzer model of calculating activity coefficients (Pitzer 1991). This approach has been used to define and measure pH values down to −3.6 for acid waters from the Richmond mine of Iron Mountain, California (Nordstrom and Alpers 1999; Nordstrom et al. 2000). Geochemical codes are available to calculate speciation in concentrated solutions using the Pitzer model, but, unlike most codes that use the ion-association model, not all of the necessary parameters for acid-sulfate waters are available. This issue has been the subject of research (e.g., Ptacek and Blowes 2000).

MINERAL ANALYSES

Numerous techniques are available to quantify the mineralogical composition and solid-phase chemical composition of tailings, waste rock, sediments, and other mineralized and nonmineralized materials. It is beyond the scope of this handbook to detail these techniques, but several good reviews are available. For instance, Crock et al. (1999) review conventional analytical techniques for solid samples, including biological samples and water samples, and point out the importance of standard reference samples. A thorough description of the mineralogy of mine wastes, including the geochemical processes and the relation of mineralogy to static and kinetic tests, can be found in Jambor (2003) and the recent Acid Drainage Technology Initiative–Metal Mining Sector handbook, *Techniques for Predicting Metal Mining Infuenced Water* (Williams and Diehl 2014). Applications of the Rietveld method for X-ray diffractometry have been described in the same volume by Raudsepp and Pani (2003). Techniques such as EXAFS (extended X-ray absorption fine structure) spectroscopy, XANES (X-ray absorption near-edge spectroscopy), micro-X-ray diffraction or X-ray microdiffraction, and X-ray microfluorescence are becoming more common and are particularly useful for identifying mine waste particles and surface bonding characteristics of important contaminants. Weisener (2003) provides a brief introduction to these techniques, whereas a more comprehensive overview on synchrotron applications is presented in the Reviews in Mineralogy and Geochemistry series (Fenter et al. 2002). Quantitative Evaluation of Minerals by SCANning electron microscopy (QEMSCAN; Butcher et al. 2000; Pirrie et al. 2004; Anderson et al. 2014) is finding wider application and allows not only for mineral identification but also mapping of elemental composition, sample surface, textural properties and contextual information, such as particle and mineral grain size and shape, mineral associations, mineral liberation and occlusion, and matrix density.

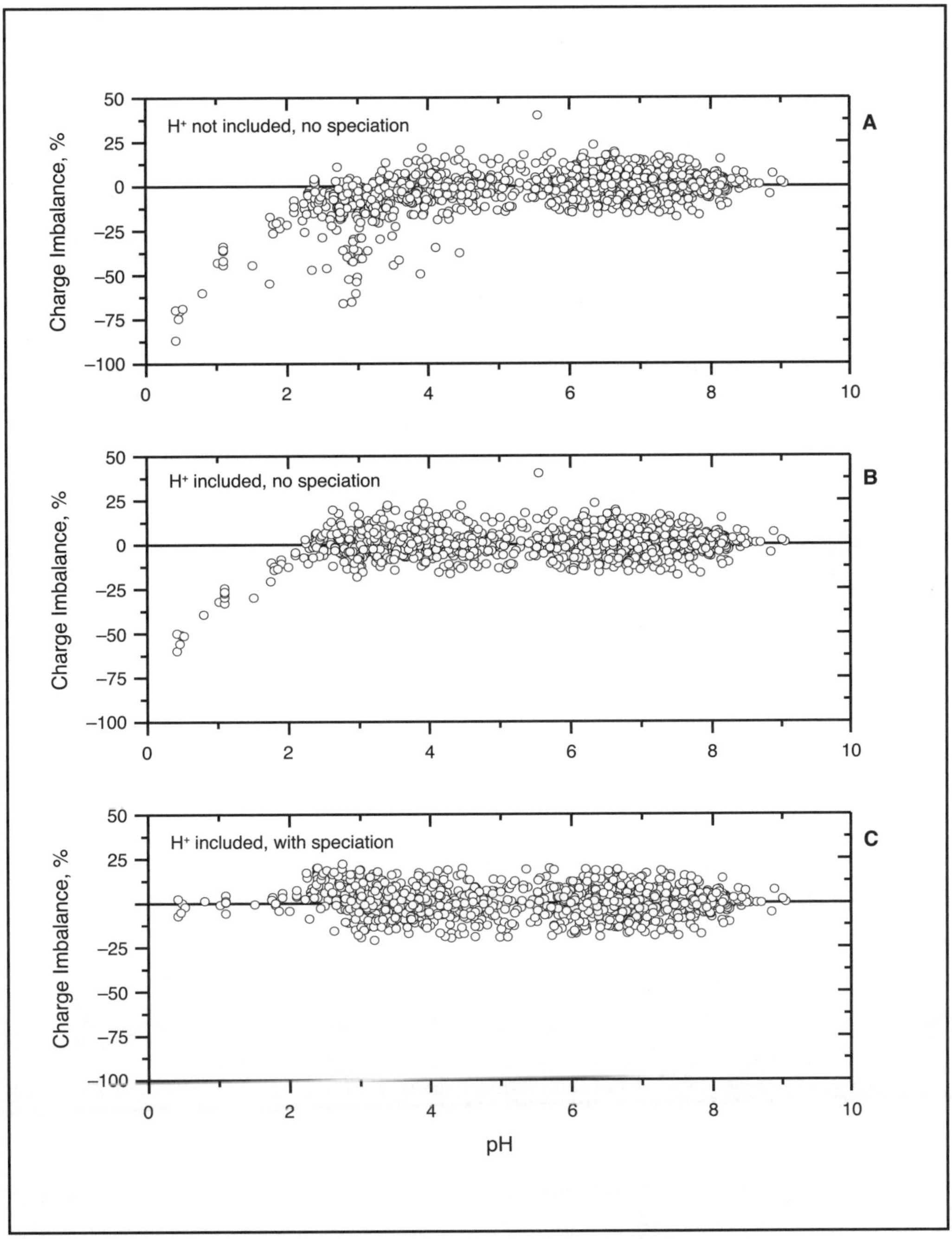

Courtesy of U.S. Geological Survey

FIGURE 5.1 (a) Charge imbalance distribution without H^+ for 1,300 mine-influenced water analyses covering a range of pH conditions. (b) Same charge imbalance distribution as (a) with H^+ but not speciated. (c) Same charge imbalance as (b) but with speciation.

MAJOR HYDROGEOCHEMICAL PROCESSES AT MINE SITES

Sulfide-Mineral Oxidation

Pyrite oxidation is the primary source of acidity from mining wastes and portal drainages. At some mine sites, pyrrhotite can be a dominant acid-producing mineral and, for some coal mines, marcasite can be important. The reactions that produce acid mine drainage (AMD) from pyrite oxidation are complex and have been studied by numerous researchers for more than a century. Stokes (1901) states, "It has long been known that ferric salts attack pyrite." The ancient civilizations of Greece, Rome, Persia, and India were familiar with pyrite oxidation and used the products (ferrous sulfate or green vitriol, and sulfuric acid or oil of vitriol) for medicine, alchemy, leaching, and blackening leather (Caley and Richards 1956; Karpenko and Norris 2002).

The main ingredients for this reaction are pyrite, oxygen from the air, and water. The reaction is catalyzed by microbes, both bacteria and archaea. A complete description of the process is beyond the scope of this handbook. Therefore, just a summary overview within the context of geochemical modeling is provided here. More comprehensive reviews can be found in, for instance, Kuang et al. (2013), Schippers et al. (2010), Baker and Banfield (2003) for the microbiology; and Chandra and Gerson (2010), Murphy and Strongin (2009), Schippers (2004), Jørgensen and Nelson (2004), Rimstidt and Vaughan (2003), Blowes et al. (2014), Younger et al. (2002), Younger and Robins (2002), Nordstrom and Alpers (1999) and numerous references therein for hydrogeology and geochemistry.

Pyrite oxidation: The chemical and graphical model. First, the stoichiometric equation for pyrite oxidation to an acid ferrous sulfate solution (only) is shown by Equation 5.5:

$$FeS_{2(s)} + \frac{7}{2}O_{2(g)} + H_2O_{(l)} \rightarrow Fe^{2+}_{(aq)} + 2SO^{2-}_{4(aq)} + 2H^+_{(aq)} \qquad \text{(EQ 5.5)}$$

This equation shows that 2 moles of proton acidity are produced for every mole of pyrite oxidized. Basically, this acidity is derived from oxidation of the sulfur moiety only and this oxidation happens very rapidly. Further oxidation of the dissolved iron is possible and occurs more slowly.

The next reaction, catalyzed by iron-oxidizing bacteria and archaea, is oxidation of the dissolved ferrous iron:

$$Fe^{2+} + \frac{1}{4}O_2 + H^+ \rightarrow Fe^{3+} + \frac{1}{2}H_2O \qquad \text{(EQ 5.6)}$$

which consumes a proton for every mole of iron oxidized. But dissolved ferric iron has a strong tendency to hydrolyze in solution:

$$Fe^{3+} + 2H_2O \rightarrow Fe(OH)_2^+ + 2H^+ \qquad \text{(EQ 5.7)}$$

and may produce 1, 2, or 3 moles of protons from every mole of ferric iron hydrolyzed depending on the pH. Combining these three reactions (Equations 5.5, 5.6, and 5.7) results in a second stoichiometric reaction for pyrite oxidation by oxygen:

$$FeS_{2(s)} + \frac{15}{4}O_{2(g)} + \frac{5}{2}H_2O_{(l)} \rightarrow Fe(OH)^+_{2(aq)} + 2SO^{2-}_{4(aq)} + 3H^+_{(aq)} \qquad \text{(EQ 5.8)}$$

which shows 3 moles of proton acidity produced from every mole of pyrite oxidized.

Finally, the hydrolyzed ferric iron can precipitate. This precipitate is often symbolized by $Fe(OH)_3$ even though such an exact stoichiometry never occurs in the field. This reaction gives us our third stoichiometry:

$$FeS_{2(s)} + \frac{15}{4}O_{2(g)} + \frac{7}{2}H_2O_{(l)} \rightarrow Fe(OH)_{3(s)} \downarrow + 2SO^{2-}_{4(aq)} + 4H^{+}_{(aq)} \qquad \text{(EQ 5.9)}$$

All of these reactions are important because rates of dissolved iron oxidation can vary considerably depending on pH, temperature, and available nutrient sources for microorganisms. Also, if the water is still in contact with pyrite, the ferrous iron will stay as ferrous; the pyrite keeps the iron reduced as rapidly as the microorganisms can oxidize the ferrous iron.

One more reaction is important to portray: the precipitation of jarosite, $KFe_3(SO_4)_2(OH)_6$, a common precipitate once the pH is less than 2.5:

$$3FeS_2 + K^+ + \frac{15}{2}H_2O + \frac{45}{4}O_2 \rightarrow KFe_3(SO_4)_2(OH)_6 + 9H^+ + 4SO_4^{2-} \qquad \text{(EQ 5.10)}$$

This reaction is not dependent on the amount of ferric iron hydrolyzed and always generates 3 moles of proton acidity for every mole of pyrite oxidized.

The previous four pyrite oxidation reactions (Equations 5.5, 5.8–5.10), and one more in which $Fe(OH)_3$ is replaced with goethite, FeOOH, can be drawn graphically by simulating the pH change as a function of the amount of pyrite oxidized. The results of this simulation, conducted with the PHREEQC code, are shown in Figure 5.2. The x-axis designates the amount of pyrite oxidized on a log scale so that the individual lines are easier to distinguish. Equation 5.5 is shown by the solid black line with a slight curvature from pH values above 4 to less than 1. Equation 5.8 is shown by the solid gray line, which at high pH values begins below the solid black line because it produces 3 protons compared to 2 protons in Equation 5.5 for every mole of pyrite oxidized. However, the gray line crosses over the black line at a pH of 3.26 because at pH values near or below the pK_1 of 2.2 for the first hydrolysis constant of Fe(III), there is little to no hydrolysis of iron and only proton consumption occurs (Equation 5.6). With precipitation of $Fe(OH)_3$, another proton is produced so that this dashed black line will be below the other lines at higher pH but will still cross over the solid black line of Equation 5.5 and at a lower pH of 2.39. If only jarosite is allowed to precipitate, the pH will follow the dotted gray line, and if crystalline goethite precipitates, the pH will follow the dotted black line. The crossover pH reflects the balance between the proton-consuming and proton-producing reactions, and indicates a minor buffering represented by the small plateau or inflection point in the curves.

From Figure 5.2 it could be estimated that if the only mineral reaction were pyrite dissolution, the resultant pH could be anywhere from 2.4 to 3.3, depending on the degree of dissolved iron oxidation and precipitation for 1 mmol of pyrite oxidized. Alternatively, for a pH of 2.0, anywhere from 3 to 16 mmol of pyrite may have oxidized, depending on the degree of iron oxidation and precipitation.

This diagram, although explaining some of the complexities of the chemistry of ARD, does not include the consequences of acid dissolution when carbonate and aluminosilicate minerals are part of the mineral suite. Nevertheless, some water analyses are available from mine sites that are predominantly affected by pyrite oxidation and little else. These samples, collected from the Leviathan mine area, California, and Iron Mountain, California (Ball and Nordstrom 1989;

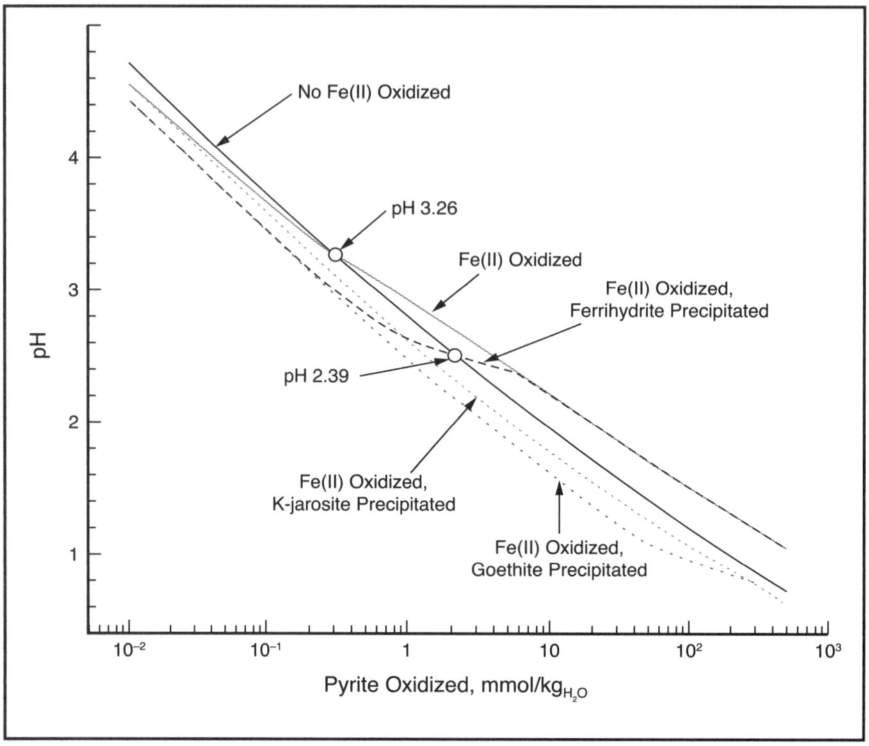

Adapted from Nordstrom and Campbell 2014

FIGURE 5.2 Simulations of pyrite oxidation under conditions in which aqueous Fe(II) is not oxidized (solid black line, Equation 3.5), aqueous Fe(II) is oxidized without precipitation (solid gray line, Equation 5.8), aqueous Fe(II) is oxidized with precipitation of ferrihydrite (dashed black line, Equation 5.9), aqueous Fe(II) is oxidized with precipitation of K-jarosite (dotted gray line, Equation 5.10), and aqueous Fe(II) is oxidized with precipitation of goethite (dotted black line, Equation 5.9) with substitution of crystalline FeOOH for $Fe(OH)_3$.

Nordstrom 1977), include a pH measured in situ and a pH measured some weeks later after the ferrous iron had oxidized and precipitated. Figure 5.3 is a reproduction of Figure 5.2 with only two lines shown for pyrite oxidation: Equation 5.5 with no ferrous iron oxidation (gray line) and Equation 5.8 with precipitation of ferrihydrite (or the most soluble hydrous ferric oxide; black line). Because there are low concentrations of cations in these samples, the initial measured pH values (open circles) can be assumed to be caused by pyrite oxidation alone. The final measured pH after oxidation and precipitation is shown by the closed circles and can be seen to closely approximate the black line in agreement with the simulation. In this comparison of a simulation with actual field data, two aspects are noteworthy. First is the good agreement, and second, because this simulation is sensitive to the chosen K_{sp} of the precipitating phase, the agreement also indicates that the solubility product constant for freshly precipitating hydrous ferric oxide is a reasonable model choice.

Carbonate Mineral Dissolution and Precipitation

Calcite ($CaCO_3$) is a very widespread carbonate mineral found in many mineralized zones as well as in many common rock and sediment types. It is an acid-neutralizing mineral that can

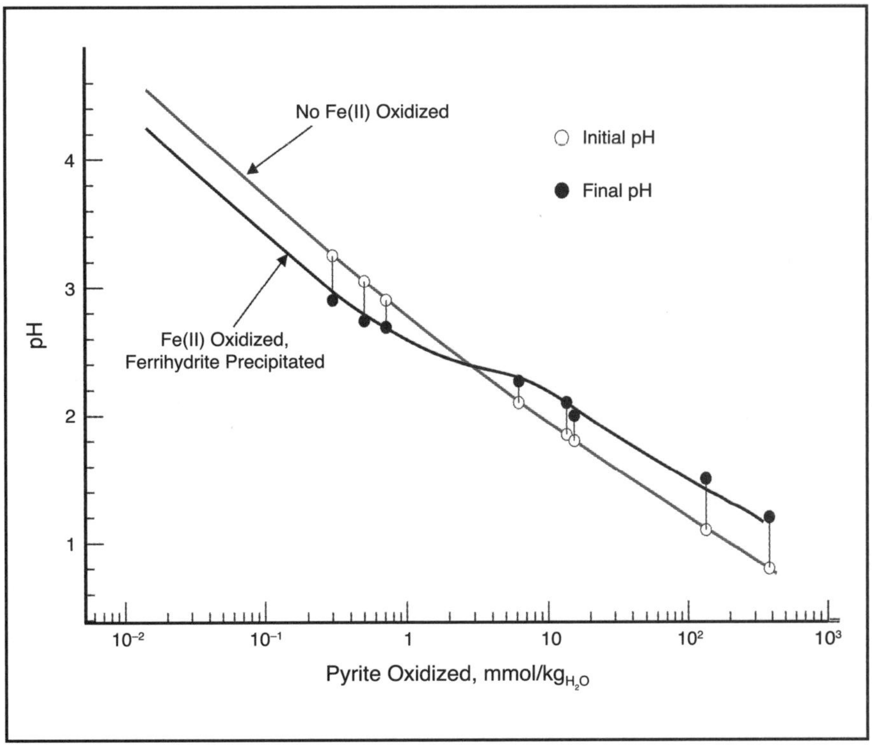

Source: Nordstrom and Campbell 2014

FIGURE 5.3 Using the lines from Figure 5.2 for just pyrite oxidation and no aqueous Fe(II) oxidation (gray line) and for pyrite oxidation with ferrihydrite precipitation (black line) (dissolved iron was predominantly Fe(II) and the sulfate was predominantly from pyrite oxidation), open circles represent measured initial field pH values while closed circles represent the final pH values of the same samples several weeks to months later after the dissolved iron had fully oxidized

counteract the acidifying effects of pyrite oxidation, and it has been used in various treatment methods. Upon dissolution

$$CaCO_{3(s)} \leftrightarrow Ca^{2+}_{(aq)} + CO^{2-}_{3(aq)} \quad \text{(EQ 5.11)}$$

it provides buffering capacity through carbonate ion hydrolysis:

$$CO^{2-}_{3(aq)} + 2H^{+}_{(aq)} \leftrightarrow HCO^{-}_{3(aq)} + H^{+}_{(aq)} \leftrightarrow H_2CO_{3(aq)} \quad \text{(EQ 5.12)}$$

Calcite dissolution in water that is in equilibrium with atmospheric CO_2 ($P_{CO_2} = 10^{-3.5}$) will dissolve up to about 0.5 mM (millimolar) or 20 mg/L of calcium at 25°C (77°F; Langmuir 1997; Nordstrom and Campbell 2014). Higher concentrations can only be obtained in acid solutions. Soils containing organic matter can produce high concentrations of CO_2, which produces carbonic acid. Under those conditions, the P_{CO_2} can reach a tenth of atmospheric concentrations and sometimes higher, while calcium concentrations can reach up to 4 mM or about 160 mg/L. Higher concentrations would require a stronger acid, such as sulfuric acid from ARD, or the presence of a more soluble source mineral, such as gypsum. Nordstrom and Campbell (2014) have simulated the effect on pH from dissolving different proportions of calcite with pyrite. Figure 5.4

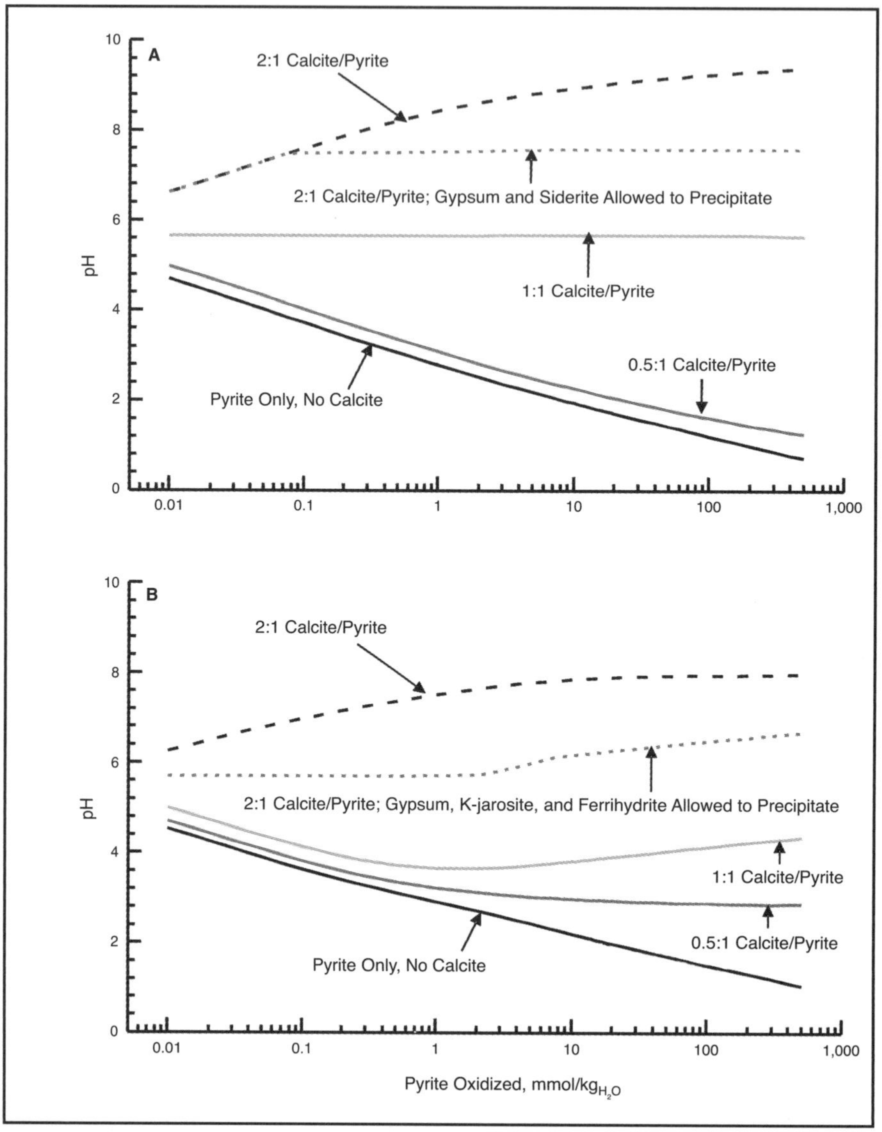

Source: Nordstrom and Campbell 2014

FIGURE 5.4 (a) Effect on pH of pyrite oxidation in the presence of different molar proportions of calcite to pyrite, without aqueous Fe(II) oxidation and without mineral precipitation (uppermost line) and with gypsum and siderite precipitation when saturation is reached (horizontal dashed line); (b) Same simulations as in (a), with Fe(II) oxidation and with and without (uppermost dashed line) the possible precipitation of gypsum, K-jarosite, and ferrihydrite

shows the modeling results for several scenarios with and without precipitation of several potential secondary minerals and with and without oxidation of the aqueous ferrous iron.

Calcite precipitation cannot occur in acid mine waters, but there are mines that have neutral- to high-pH groundwaters that precipitate calcite (Nordstrom et al. 1989; Banks et al. 2002). Tailings effluents can have high-pH values from residual flotation or leaching agents used to extract metals and may precipitate calcite on reaction with atmospheric CO_2.

Gypsum Dissolution and Formation

Gypsum, $CaSO_4 \bullet 2H_2O$ (or anhydrite, $CaSO_4$) commonly occurs as an accessory mineral in mineral deposits, formed during weathering of coexisting pyrite and calcite in mining wastes (e.g., Ludington et al. 2005; Blowes et al. 1991). By itself, gypsum dissolution or precipitation does not affect pH, but it can be a sink for the removal of sulfate from solution. Acid mine waters that are high in sulfate concentrations are often near or at saturation with respect to gypsum solubility. The calcium concentration for equilibrium solubility of gypsum is 612 mg/L while the corresponding sulfate concentration is 1,470 mg/L. Additional sulfate from pyrite oxidation is easy to recognize in a water analysis because of the molar ratio of $SO_4/Ca > 1$ in addition to the low pH. Most groundwaters have Ca/Mg molar ratios greater than or close to 1. Molar ratios of Ca/Mg can be substantially less than 1 in acid mine waters because of the solubility limit imposed by gypsum, whereas much higher Mg concentrations would be necessary to reach a solubility limit for a Mg-SO_4 salt such as epsomite.

A simple example illustrates how mass balances work with a hypothetical calculation in which only gypsum and pyrite are weathering. Given a water analysis with 480 mg/L of SO_4 (5 mM) and 50 mg/L of Ca (1.25 mM), it can be concluded that 1.25 mmol kg^{-1} of gypsum is weathering and 1.875 mmol kg^{-1} of pyrite is weathering. More commonly, calcite, gypsum (or anhydrite), and pyrite are all weathering, and to solve simultaneously that set of equations, one must know the concentrations of Ca, alkalinity, SO_4, and Fe in both redox states. Another approach that may prove beneficial is to include the sulfur-isotope composition of the dissolved sulfate. The sulfur-isotope constraint only applies if the gypsum/anhydrite is primary (i.e., formed at high temperature with the main mineralization). Under those conditions, the gypsum/anhydrite will be enriched in $\delta^{34}S$ relative to that of the pyrite (Dold and Spangenberg 2005). Nordstrom et al. (2007) showed that water samples collected from the San Juan Mountains mineralized areas tended to fall into three categories of calcite-dominated, gypsum-dominated, and pyrite-dominated water types (Figure 5.5). Samples inside the triangular distribution can be assumed to be mixtures. Figure 5.5 represents a graphical depiction of mass balances. If the gypsum formed from weathering of pyrite, then the pyrite and gypsum will have very similar isotopic compositions.

Aqueous Oxidation and Reduction of Trace Metals and Metalloids (such as Fe, Mn, As, S, Se, Sb, Cr, U, and V)

Speciation of redox-sensitive elements greatly affects their mobility and toxicity. Redox chemistry is so important and redox disequilibria are so prevalent (Lindberg and Runnels 1984; Sigg 2000) that the only reliable way to determine redox conditions is to analytically determine the individual species. Redox potential (Eh) measurements with a conductive inert electrode, such as platinum or gold, only behave quantitatively in acid mine waters when the iron concentrations are $\geq 10^{-5}$ m, which would require pH values to be ≤ 4.0 (Nordstrom 2011), or in sulfide-rich waters (R.A. Berner 1963). That is, with very careful analytical measurements and a thorough understanding of electrochemical equilibrium, it is possible to show a quantitative relation between a redox potential measurement and the Fe(II/III) redox couple in an acid water or the dissolved sulfide concentration in a sulfide-rich environment. In nearly all other situations, the redox electrode measurement will not give an equilibrium value of Eh. Even in the two conditions just noted, the equilibrium is only between a particular redox couple and the electrode surface. An equilibrium redox potential in acid mine waters only means that the Fe(II/III) redox couple is consistent with a Nernst equation redox potential measurement, not that the water as a whole is in redox equilibrium. There could be dissolved oxygen and dissolved methane in the same water as that containing the Nernstian Fe(II/III) equilibrium, which means that there is no equilibrium redox potential

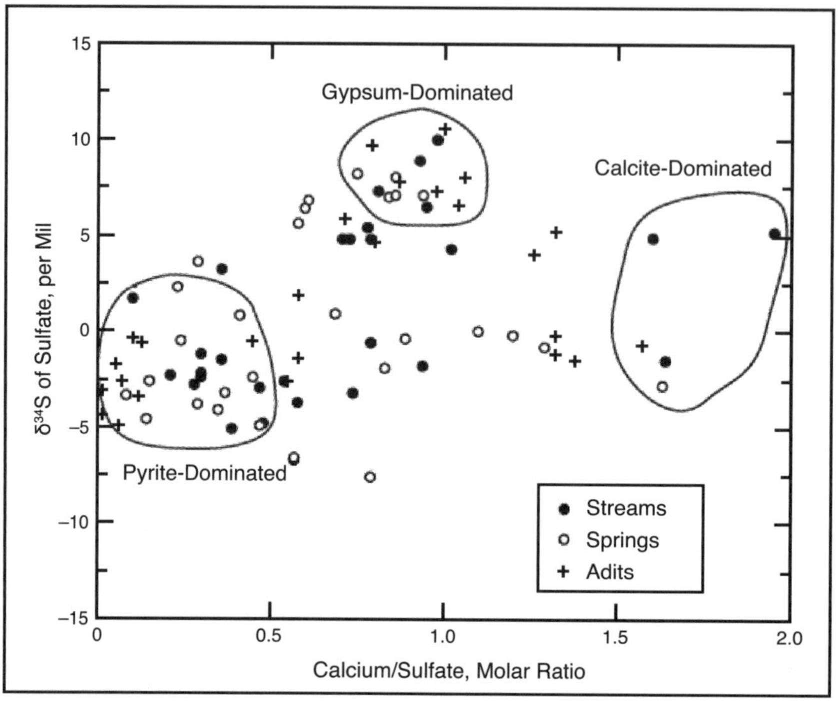

Source: Nordstrom et al. 2007

FIGURE 5.5 The $\delta^{34}S$ of dissolved sulfate in streams, springs, and mine adits plotted against the calcium/sulfate molar ratio for samples from the San Juan Mountains

for the water sample. It has been stated quite forcefully and correctly that to assign a redox potential to a water sample is meaningless (Thorstenson 1984; Hostetler 1984). Hence, redox species in natural waters cannot be modeled or simulated based on electrode measurements, and they should always be determined analytically. Table 5.1 compiles the main redox-active species that might be present at a mine site, as well as their relative solubility and toxicity as a function of their redox state. Solubility, toxicity, and aqueous mobility are strongly related to the redox state of an element. Selenium is an essential element at very low concentrations and toxic at slightly elevated concentrations. Chromium in the reduced state (III) is an essential element but only needed in very low concentrations; in the oxidized form (VI) it is highly toxic at very low concentrations. Because iron (III) oxyhydroxides are highly insoluble except at very low-pH values and because these iron precipitates are excellent scavengers of trace elements, oxidation–reduction processes that dissolve or precipitate iron will have a major control on the dissolved concentration and mobility of other trace elements (Jenne 1968).

Eh-pH diagrams. Activity diagrams known as Eh-pH, pe-pH, or Pourbaix diagrams are often used to delineate the range of redox potential and pH in which aqueous species and mineral species are stable (Hem and Cropper 1959; Garrels and Christ 1965; Brookins 1988) or metastable (Sato 1992). Their main purpose is to show the predominance areas of pH and redox conditions in which one would expect to find a particular species. In this sense, they are helpful as graphical models. Because of the inability of these diagrams to portray multiple temperatures, multiple water compositions, and complete speciation, they should generally not be used quantitatively to plot water analyses and draw interpretive conclusions. A preferable diagram for interpreting water analyses would be a saturation index plot, of which several varieties are in use.

TABLE 5.1 Some common redox-sensitive elements and their relative solubilities and toxicities

Element	Reduced Form	Oxidized Form
Fe	Fe(II), soluble	Fe(III), insoluble
Mn	Mn(II), soluble	Mn(IV), insoluble
As	As(III), soluble + more toxic	As(V), insoluble + less toxic
Se	Se(IV), insoluble + less toxic	Se(VI), soluble + more toxic
Cr	Cr(III), insoluble + less toxic	Cr(VI), soluble + more toxic
U	U(IV,V), insoluble	U(VI), soluble
V	V(III), insoluble	V(IV,V), soluble

Courtesy of U.S. Geological Survey

An example of an Eh-pH diagram for Fe is shown in Figure 5.6. This diagram is based on thermodynamic data, and the derivation of the diagram can be found in Nordstrom and Munoz (1994). Under lower pH and more oxidizing conditions, the predominant species are aqueous Fe^{2+} and Fe^{3+}, but on the right side of the diagram, the abundance of aqueous Fe is much lower because relatively insoluble minerals are stable, such as ferrihydrite and, to a lesser extent, siderite. This equilibrium diagram, therefore, not only shows the relative stabilities of minerals in terms of redox potential and pH, but it also indicates that aqueous Fe will be present in lower concentrations on the right side of the diagram and higher concentrations on the left. Hem (1972) drew a similar diagram for Mn that depicted similar characteristics, with less soluble Mn on the right side because of the formation of relatively insoluble oxides and the carbonate mineral rhodochrosite under more oxidizing and reducing conditions, respectively. The left side of the diagram for Mn shows soluble Mn^{2+} as being the predominant species. Hem then superimposed the Mn Eh-pH diagram on the Fe diagram, as reproduced in Figure 5.7. The solid line divides the less soluble Mn from the more soluble Mn, and the dashed line divides the less soluble Fe from the more soluble Fe. Although they have somewhat similar characteristics, Hem notes that if a water of fairly constant pH near 7 becomes more reducing, Mn reduces before Fe. Hence, as a groundwater goes anoxic, Mn concentrations should increase before Fe concentrations because one would cross the solid line in Figure 5.7 first and then the dashed line. Conversely, if groundwater encounters more oxidizing conditions, Fe will oxidize and form precipitates first, followed by Mn, resulting in the brown and black banding sometimes encountered along streambanks at mining sites. These diagrams are helpful to clarify some pH and redox differences between redox-sensitive elements as well as those conditions that promote high or low concentrations of these elements.

Iron Mineral Precipitation

Acid mine waters mobilize large quantities of iron from pyrite oxidation. If the iron is kept reduced, it can only precipitate as siderite, $FeCO_3$, or as an efflorescent salt such as melanterite, $FeSO_4 \bullet 7H_2O$. Upon oxidation, iron precipitates, often as a mixture of different mineral phases. Initial precipitates are usually amorphous to normal X-ray diffractometry and they transform with ageing (Kumpulainen et al. 2008). These colloidal precipitates go by many different names, the most common being HFOs (hydrous ferric oxides; Dzomback and Morel 1990) and ferric oxyhydroxides (Whittemore and Langmuir 1975). With improved techniques, it has been possible to identify these initial precipitates and their aged products as ferrihydrite, $Fe_5HO_8 \bullet 4H_2O$; schwertmannite, $Fe_8O_8(OH)_6SO_4$; goethite, FeOOH; and jarosite, $KFe_3(SO_4)_2(OH)_6$ (Kumpulainen et al. 2008; Bigham and Nordstrom 2000; Cornell and Schwertmann 2006). Their chemical

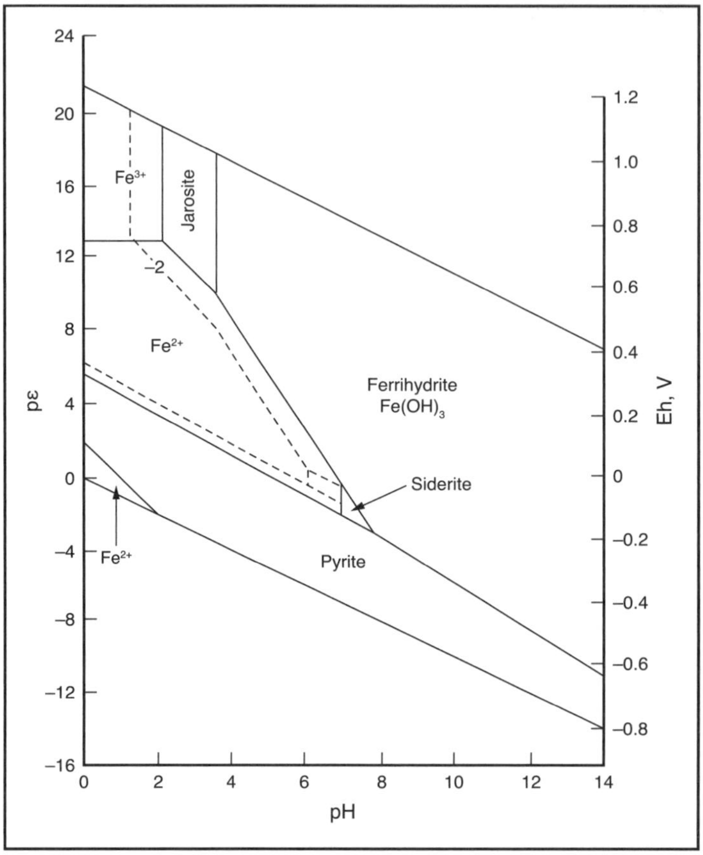

Adapted from Nordstrom and Munoz 1994

FIGURE 5.6 Eh-pH (and pε-pH) diagram for Fe at 25°C (77°F), $\Sigma Fe = 10^{-4}$ m (dashed line at 10^{-2} m), $\Sigma S = 10^{-2}$ m, and $P_{CO_2} = 10^{-2}$ bar

compositions are highly variable, depending on the amount of substitution of metal cations or hydronium ion, H_3O^+, for Fe or K, and of anions such as AsO_4^{3-}, Cl^-, F^-, and PO_4^{3-} for OH^-.

Two important aspects to modeling iron precipitation are (1) the equilibrium speciation and solubility and (2) the rate of oxidation and precipitation. The speciation modeling depends on the reliability of the thermodynamic stability constants and activity coefficients and their dependence on temperature, ionic strength, and water composition. Rates of oxidation can be highly variable and dependent on additional factors, especially microbial growth rates. Recently, a kinetic model for Fe oxidation and precipitation was developed and applied to pipe scale formation in a pipe carrying AMD (Campbell et al. 2013). An example of Fe oxidation with accompanying changes in pH is shown in Figure 5.8. During aqueous Fe oxidation by *Acidithiobacillus ferrooxidans*, the pH increases at first. Later, the acid-producing hydrolysis reaction brings the pH down, and finally precipitation of Fe minerals brings the pH down even further while essentially eliminating Fe from solution. Rates of Fe oxidation, hydrolysis, and precipitation all have to be treated differently.

The precipitation of hydrated ferric oxides is common in AMD and the saturation indices (SIs) for $Fe(OH)_3$ (similar to ferrihydrite) reflect an equilibrium solubility when calculated based on careful determinations of Fe(II/III) and the major ions in the water sample so that speciation

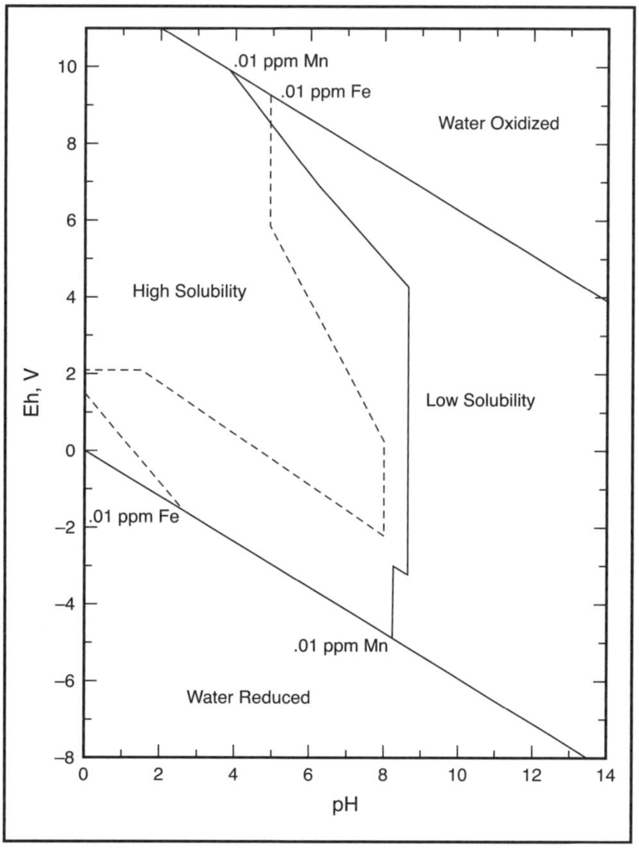

Source: Hem 1972

FIGURE 5.7 Combined and simplified Eh-pH diagram for both Fe (dashed line) and Mn (solid line), which tends to separate those conditions that promote high or low concentrations of these elements

is possible. Figure 5.9 is an example of carefully collected and analyzed data from the Questa project in New Mexico (see chapter 9 case studies). Here the SIs of ferrihydrite are plotted as a function of the iron concentration (Nordstrom 2008). The SIs are all fairly constant for Fe concentrations above about 10 mg/L (2×10^{-4} M), suggesting an equilibrium threshold had been reached between freshly precipitated Fe ($\log K = 4.9$) and microcrystalline Fe ($\log K = 3.0$). One of the challenges with trying to interpret saturation indices for hydrated ferric oxides is that fresh precipitates tend to be in the nanocolloid range and do not filter well. Nordstrom (2011) demonstrated how this problem could be discerned by careful application of detection limits for dissolved Fe(III) and for redox potential measurements. For pH values higher than about 4, the Fe concentration in a water sample can have substantially more colloidal Fe than truly dissolved Fe(III), and much of these colloids can penetrate filters and become included in the "dissolved" fraction.

Aluminosilicate Mineral Dissolution and Precipitation

Feldspars, micas, clay minerals, chlorites, and epidotes are common aluminosilicate minerals that often accompany ore minerals and may, therefore, appear in waste rock and tailings. Although they have some capacity for neutralizing acid waters, their ability to do so is limited. Their

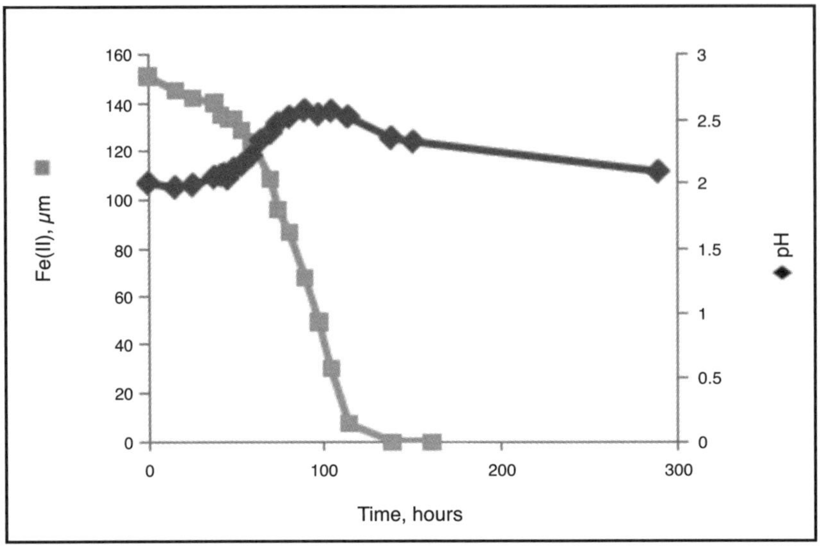

Source: Campbell et al. 2012

FIGURE 5.8 Fe oxidation under laboratory experimental conditions showing concurrent changes in pH. Changes in pH reflect the different sequence of reactions from oxidation to hydrolysis to precipitation.

dissolution rate is slower than that for carbonates, and they can acquire silica-rich surface layers that inhibit further dissolution. Feldspars, micas, chlorites, and epidotes tend to dissolve under acidic conditions to form clay minerals such as kaolinite. The following reaction is an example of the acid dissolution of albite to form kaolinite:

$$2NaAlSi_3O_8 + 2H^+_{(aq)} + 9H_2O \rightarrow Al_2Si_2O_5(OH)_4 + 2Na^+_{(aq)} + 4H_4SiO_{4(aq)} \quad \text{(EQ 5.13)}$$

With increasing sulfuric acid acidity, kaolinite will break down to form alunite:

$$3Al_2Si_2O_5(OH)_4 + 2K^+_{(aq)} + 6H^+_{(aq)} + 4SO^{2-}_{4(aq)} + 9H_2O \rightarrow 2KAl_3(SO_4)_2(OH)_6 + 6H_4SiO_{4(aq)} \quad \text{(EQ 5.14)}$$

Both alunite and kaolinite can also form under hydrothermal conditions so that careful mineralogical examination would be necessary to distinguish a hypogene (hydrothermal) clay from a supergene (weathered) clay. This has been a classic controversy about how to distinguish supergene from hypogene mineral formation. It can usually, but not always, be resolved through the use of isotopic determinations. The reaction of other aluminosilicate minerals with sulfuric acid can be written similarly and the overall process is the same: alkaline and alkaline earth metals will be released to solution and a clay will form with excess silica going into solution, which should form some silica phase (amorphous silica or microcrystalline quartz) mixed in with the clays. These reactions can be simulated with several computer codes, and examples can be found in Appelo and Postma (2007) and Merkel and Planer-Friedrich (2008).

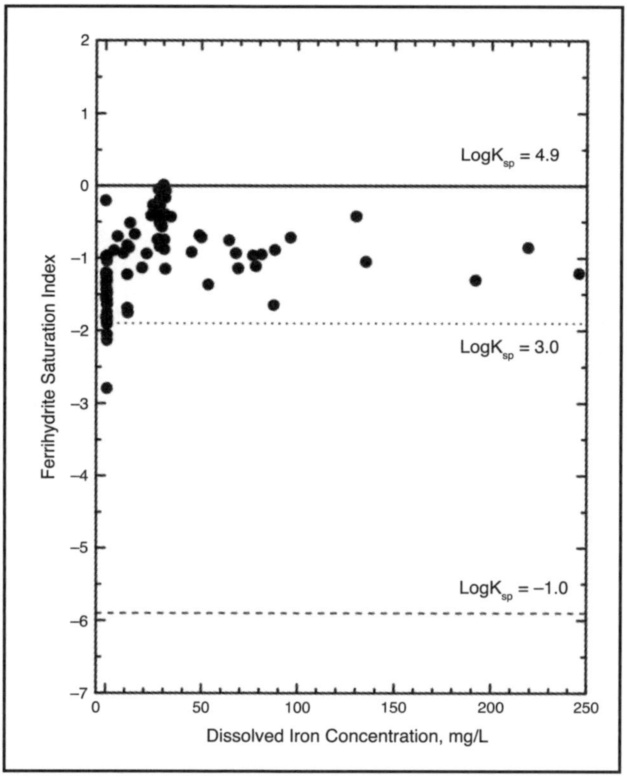

Source: Nordstrom 2008

FIGURE 5.9 Ferrihydrite saturation indices for Fe concentrations up to nearly 250 mg/L with relatively constant values in the range of freshly precipitating hydrated ferric oxide from laboratory studies

Aluminum and Silica Precipitation

Acid waters have elevated Al and SiO_2 concentrations because most aluminosilicate minerals dissolve more readily at pH values near 3 or less. Acid mine waters typically have Fe and Al as the two major cations. As these waters migrate downgradient in surface waters or groundwaters, the Al will precipitate when the pH begins to approach the pK of the first hydrolysis constant (Nordstrom and Ball 1986), which is 5 (Nordstrom and May 1996). At lower pH values than 4.5 for surface waters and at pH values lower than 4 for groundwaters (Blowes et al. 2014; Nordstrom 2008), Al behaves like a conservative constituent. It begins precipitating at pH 4 to 5, usually as an "amorphous" basaluminite ($Al_4(SO_4)(OH)_{10}\bullet 5H_2O$; Bigham and Nordstrom 2000). At pH values higher than about 7, the dissolved organic carbon is often present at higher concentrations than the Al, and Al-organic complexes dominate. If Al-organic complexing is not part of the code calculation, then a downward trend, in the SI for aluminum hydroxide with pH, will naturally occur because the code assumes an increasing amount of the $Al(OH)_4^-$ species with increasing pH. That effect decreases the activity of Al^{3+} and the SI, as shown in Figure 5.10.

Another clay that commonly forms in waters, soils, and sediments over a large range of pH is gibbsite, $Al(OH)_3$. Two other aluminous clays, diaspore and boehmite, both AlOOH, are also found as weathering products, but not as commonly as gibbsite. In acid-weathering conditions when silica concentrations are not high enough to form kaolinite, gibbsite can form. However, in acid mine waters, basaluminite and hydrobasaluminite, $Al_4(SO_4)(OH)_{10}\bullet 12–36H_2O$, are more

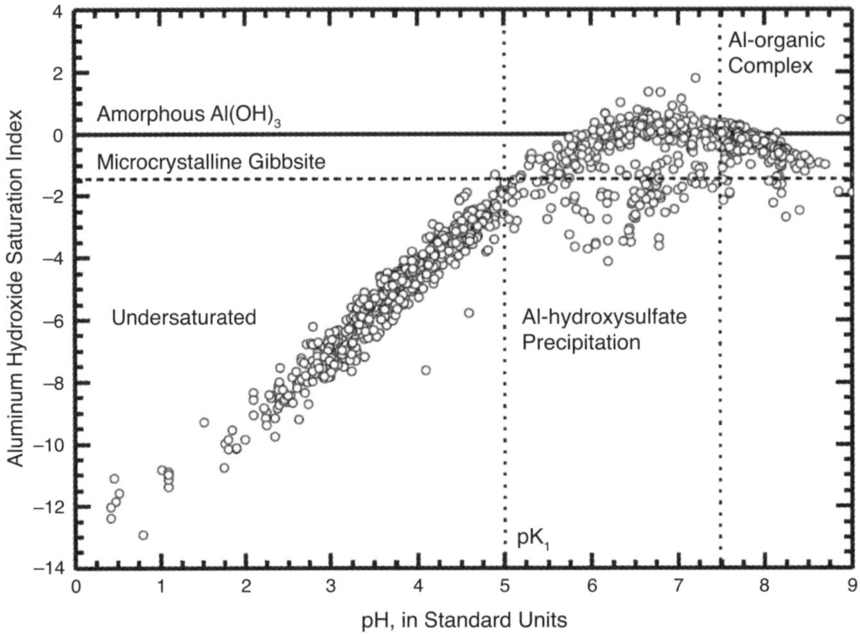

Note: Three distinct regions can be observed: a region of undersaturation for aluminous minerals generally below about 5, an approach to saturation limits at pH values of 5–7.5, and a downward trend for pH >7.5 that is an artifact of speciation modeling when organic complexing is not accounted for.

Source: Nordstrom 2011

FIGURE 5.10 Saturation indices for $Al(OH)_3$ relative to pH for more than 1,300 surface water samples from the western United States

likely the precipitating phases. These precipitates are often observed as white, gelatinous material or white coatings on rock surfaces where an acid water has mixed with a neutral water and raised the pH to about 4.5 or higher (Nordstrom and Ball 1986; Bigham and Nordstrom 2000). Many observations have been made of white precipitates in mine waters, and they have always been where mixing occurs with a consistent pH of 5 to 5.5 after mixing. The two photos in Figure 5.11 are typical examples. Bigham and Nordstrom (2000) review examples of aluminous minerals precipitating in mine effluent and their conditions of formation.

Efflorescent Salt Formation and Dissolution

Soluble sulfate salts are common on waste rock, tailings piles, and unmined outcrops of sulfide minerals. They form a frosty white or colored coating and reflect the water chemistry from which they effloresced during drying of mine waters. Thermodynamic properties for these salts and their associated solutions have not been fully worked out so that geochemical modeling is limited. The solutions that they form from are very concentrated and require the use of the Pitzer model and, in some cases, new experimental measurements. These salts should be thought of as dried AMD, stored in solid form. They quickly dissolve when the next rainfall or snowmelt event occurs and may contribute metals and acid to the receiving waters. Keith et al. (2001) modeled the change

Courtesy of U.S. Geological Survey

FIGURE 5.11 (a) White aluminous precipitate approximating the composition of basaluminite but amorphous to X-rays in effluent from a caved-in mine portal near Silverton, Colorado; (b) white aluminous precipitate in Leviathan Creek, near Markleeville, California

in composition of Boulder Creek (California) with different mass amounts of different salts dissolving in the runoff from a rain event using the EQ6 code (Figure 5.12). These salts are all very soluble and often found in and on mine wastes with oxidizing sulfides. With the one exception of hexahydrite, a magnesium sulfate hydrate with no hydrolysable cation, all of these salts produce low-pH solutions when dissolved.

Mixing of Waters with Reaction

As AMD empties from a mine site, it will inevitably mix with a stream or river of neutral pH and having some neutralization capacity. If the pH changes enough to cause supersaturation with respect to a mineral, then it will precipitate. The other situation is oxidation of dissolved Fe(II) to an insoluble iron precipitate that will sorb some metals. Iron staining is common on stream sediments that are exposed to AMD. The rate of the reaction depends primarily on the density of iron-oxidizing microbes but also on pH, temperature, and the concentration of iron. Another confounding aspect is photoreduction whereby dissolved Fe(III) is reduced by sunlight during the day and reoxidized by microbial communities during the night. Increases in pH from mixing increases the oxidation rate. Modeling the rate of iron oxidation is complicated, but several investigators have been successful in laboratory and field studies (Kirby and Elder Brady 1998; Kirby et al. 1999; Geroni and Sapsford 2011; Chen and Jiang 2012; Campbell et al. 2013). Once dissolved iron is oxidized, it will precipitate at pH values at or above 2. Aluminum precipitates at pH values of 4–5 or higher. These reactions are so consistent that it is possible to estimate the pH from what is seen to precipitate in the water. The reason they are so consistent is because the precipitation is a hydrolysis reaction that depends on the pK_1 value of the first hydrolysis constant for the metal involved. The pK_1 for Fe(III) is about 2 and the pK_1 for Al is 5. The relation between this pK_1 for

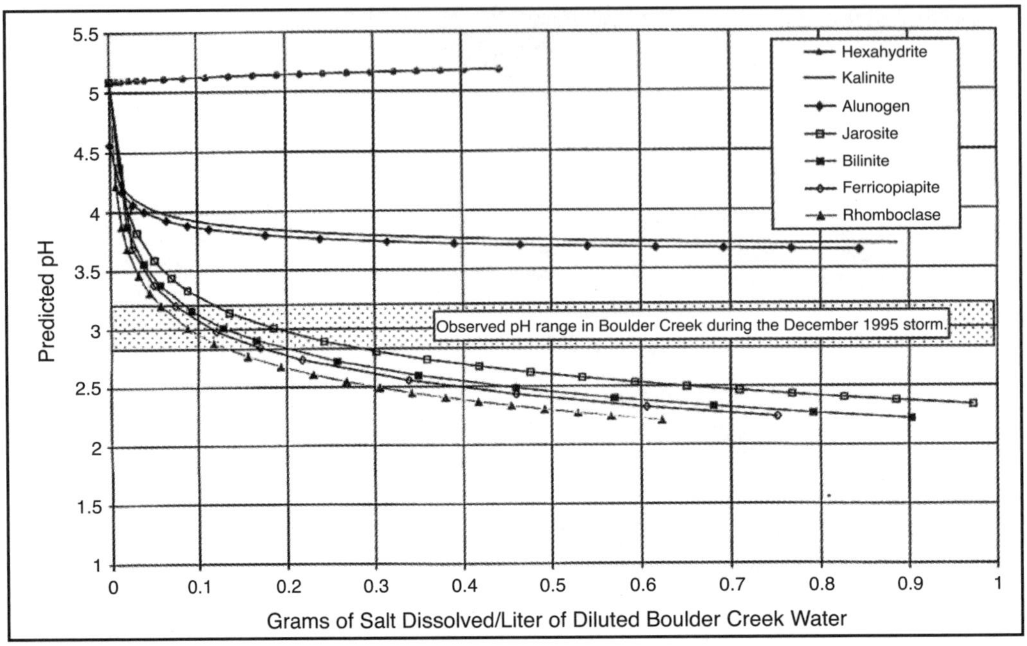

Source: Keith et al. 2001

FIGURE 5.12 Plot showing the effect of different amounts of efflorescent salt dissolution on the resultant pH of Boulder Creek water (with reference to the Iron Mountain Superfund site)

Al and the concentration of dissolved Al as the pH increases from mixing was shown clearly in field data (Nordstrom and Ball 1986) as mentioned previously (Figure 5.11).

Evaporation of Waters with Reaction

Evaporation of AMD increases the concentration of dissolved solutes and lowers the pH. Minerals eventually precipitate from the solution because saturation is reached. The minerals that precipitate at low pH (<1) are usually soluble salts and efflorescences (mentioned previously). The mineralogy of these salts has been well characterized and their rapid dissolution during rainstorm events has also been documented (Jambor et al. 2000; Nordstrom 2009, 2011). This process of evaporation and soluble salt formation is common at most mine and mineral processing sites, but in underground mines it is often enhanced by the higher temperatures encountered where pyrite is actively oxidizing. Pyrite oxidation is a strongly exothermic reaction, and the extreme case is at Iron Mountain, California, where underground temperatures reached 221°C (430°F; Wright 1906) and the massive sulfide became incandescent. Massive quantities of soluble salts are found underground at Iron Mountain and are thought to be the main control on the chemistry of the mine effluent (Nordstrom and Alpers 1999).

Modeling the effect of rainstorm dissolution of soluble salts is challenging because it partly depends on the intensity and duration of the rainstorm, and the accumulation of salts that is dependent on the sulfides oxidizing, the microbial community, and the duration of the dry period before the rainstorm. Just modeling the water flux alone during rainstorm events is extremely complex. This phenomenon is highly dynamic and transient, but the more careful field studies can be that can capture the variations in water chemistry through these events, the better we can anticipate the future events.

CHAPTER 6

Hydrologic Modeling

Walter Weinig

INTRODUCTION

Impacts due to acid drainage occur primarily in aqueous environments. Consequently, hydrologic models are important components of the general geochemical modeling process. Hydrologic models can be used to provide a volumetric flux estimate, for example, in problems related to groundwater impacts. They can also provide estimates of mixing volumes, often needed in problems related to pit lakes and more complex hydrogeologic settings.

The science of hydrology encompasses surface water, groundwater, and the vadose zone. Surface water includes water encountered at the earth's surface, such as streams, rivers, and lakes. *Groundwater* generally refers to water contained within water-saturated geologic media. The vadose zone includes unsaturated geologic media, most often encountered between the ground surface and the zone of full saturation.

The type of model output needed and the critical elements identified in the site conceptual model will drive the selection of a suitable, quantitative hydrologic model. Hydrologic models may be designed to represent an integrated surface and subsurface system, or may represent only a portion of the full system, depending on the particular problem at hand. For complex systems or to evaluate the importance of different physicochemical processes, multiple hydrological models may be applied, either separately or coupled, to predict impacts due to acid drainage.

This chapter is divided into sections describing modeling approaches for major portions of the water cycle that are relevant to acid drainage problems. Water balance models are sometimes used to quantify a conceptual water cycle at the site scale without explicitly representing the physical processes involved in each part of the cycle. The integrated water cycle provides a framework for identifying discrete portions of the overall system.

Examples of computer codes are provided to illustrate the types of approaches that are available for different hydrologic components. The codes described in this chapter represent only a subset of the programs available in the public sector and commercially. Mention of a particular code is not meant to imply its suitability for any particular problem. Similarly, lack of inclusion of a code is not meant to imply that it cannot be, or has not been, applied to acid drainage problems.

WATER CYCLE

The water cycle, or hydrologic cycle as it may also be called, is itself a conceptual model of a complex system. A generalized depiction of the water cycle is shown in Figure 6.1.

The water cycle provides a useful conceptual model to identify important processes that may impact a particular problem. For example, near-surface processes such as evaporation, transpiration, and precipitation are important for problems involving surface water and the vadose

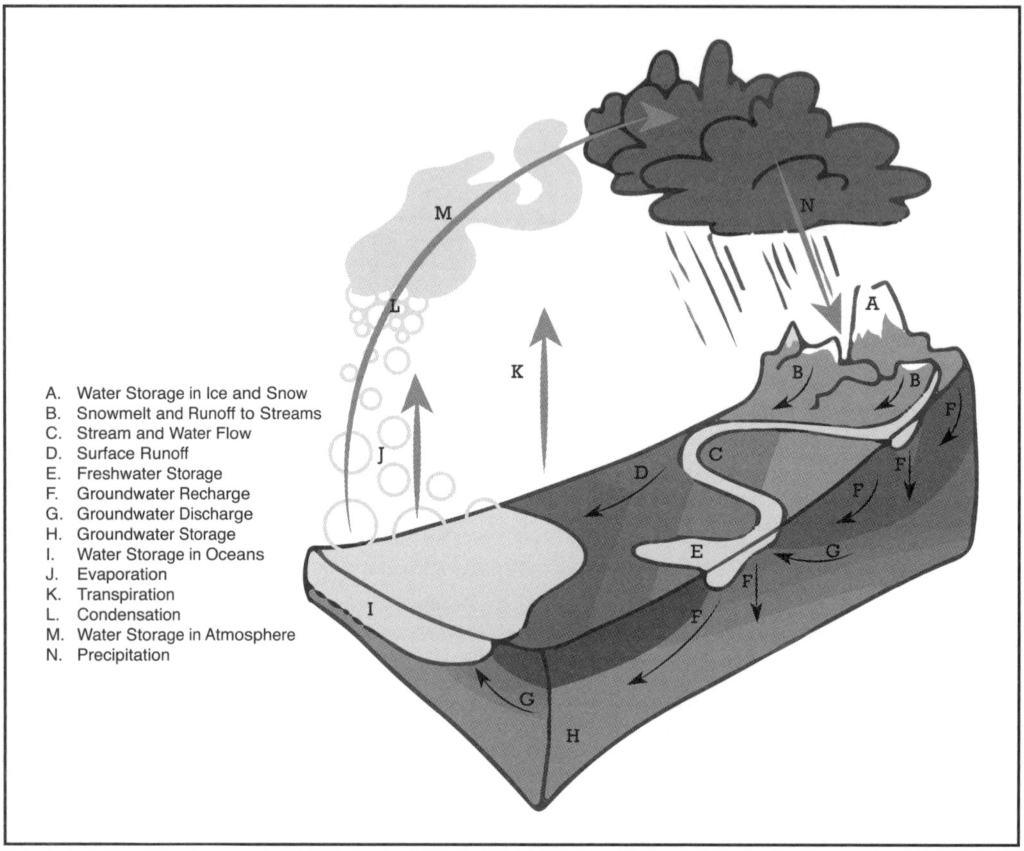

FIGURE 6.1 Generalized water cycle

zone. Infiltration and groundwater discharge may be more important for problems affecting drinking-water aquifers. In groundwater models, near-surface processes may be represented with less precision or the effects of such processes may be combined into a single parameter such as areal recharge.

WATER-BALANCE MODELS

Water-balance models are an empirical approach to quantifying a conceptual hydrologic model. The main elements of the water cycle that are applicable to the site and problem under consideration can be abstracted as reservoirs, inputs, outputs, and interconnections between elements. The water-balance model operates by quantitatively tracking inflows and outflows from and between the various components of the conceptual model, tracking mass or volume in each step, and identifying limiting capacities for volume or flux. For dynamic models, time lags for flow between elements can be incorporated. Water-balance models are types of flux-reservoir models as described in Chapter 3.

The basic equation for a water-balance model relates inflows and outflows to changes in storage over the model domain, incorporating principles of continuity and conservation of mass:

$$[\text{inflow}] - [\text{outflow}] = \Delta[\text{storage}] \qquad \text{(EQ 6.1)}$$

Inflows can include elements such as

- Direct precipitation;
- Surface-water runoff;
- Groundwater inflows;
- Process-related inflows; and
- Water contained in slurries, for example, in tailings facilities.

Outflows can include elements such as

- Evaporation from surface-water bodies or bare soils;
- Transpiration from plants;
- Surface-water runoff;
- Groundwater outflows;
- Engineered water discharges from the model domain; and
- Subsurface discharges, for example, seepage out of the base of unlined waste rock or tailings storage facilities.

Changes in storage can include elements such as

- Increases or decreases in moisture content within unsaturated materials such as waste rock or tailings,
- Increases or decreases in water levels within surface ponds or reservoirs,
- Increases or decreases in groundwater elevations in unconfined aquifers, and
- Increases or decreases in potentiometric head in confined aquifers.

Water-balance models address a variety of scales and complexities. For example, a problem involving potential impacts from a waste rock pile may involve a relatively simple water balance around that individual facility. A problem related to predictions of impacts from tailings may include the tailings storage facility, decant pond, the mill feeding the tails, and water recycling back to the mill.

A water-balance model can be developed around a natural domain such as a drainage basin or regional catchment. The domain may include several mining-related facilities. This type of domain may be appropriate if impacts are assessed at a downstream point of compliance.

A site-wide water balance for a major mining operation can be very complex, with many individual facilities and water moving in intricate cycles. The level of detail and number of connections between individual elements may make it impractical to use a site-wide water balance for geochemical assessment without appropriate simplifications to focus on the most important elements of the conceptual model.

Water-balance models can be constructed in spreadsheets. Spreadsheet approaches can be limited if the system being studied has many elements or if transient effects are important. More complex water-balance or mixing-cell models, incorporating complex physical or chemical processes, transient effects, or probabilistic components, can be addressed by using generalized dynamic system simulators such as GoldSim (Hall et al. 2006) or Stella (Wright 1979).

Mass and concentration of dissolved constituents can be incorporated into this approach through mixing-cell models, lookup tables, or by directly linking geochemical models such as PHREEQC or The Geochemist's Workbench (e.g., Usher et al. 2010; Eary 2007) to the water-balance model. Using these approaches, concentrations and masses of individual constituents can be incorporated into the model to evaluate changes in water quality over time in the system being studied.

SURFACE-WATER MODELS

Surface-water models are used to assess runoff resulting from precipitation and flow in streams or channels. Runoff represents the surface flow after infiltration and initial surface abstraction has occurred. Appendix A of *EPA and Hard Rock Mining: A Source Book for Industry in the Northwest and Alaska* (USEPA 2003) describes the basic approaches to modeling runoff processes based on precipitation inputs.

A common approach to estimate the volume of runoff is the Soil Conservation Service (SCS, now known as the Natural Resources Conservation Service or NRCS) curve-number method (Mockus 1972). The SCS curve-number method requires an evaluation of surface characteristics of a watershed, determining the resulting curve number, then applying that number along with precipitation information to calculate the resulting volume of runoff according to the following equation:

$$Q = \frac{(P - 0.2S)^2}{(P + 0.8S)} \quad \text{(EQ 6.2)}$$

where

Q = volume of direct runoff, inches over the watershed area

P = precipitation volume determined from rainfall data or rainfall-frequency assessment at the site, inches over watershed area

S = fitting parameter representing potential maximum retention after runoff begins

The maximum retention, S, is calculated as

$$S = \frac{1{,}000}{CN} - 10 \quad \text{(EQ 6.3)}$$

where CN is the curve number from tables based on land cover type and hydrologic soil group.

Estimates of runoff at a particular site may also be accomplished using the unit-hydrograph method. This method is also described in Mockus (1972). To apply the unit-hydrograph method, a hydrograph relating runoff to precipitation is developed for a unit precipitation volume over an area, for example, 1 in. or 1 cm of rainfall. This unit hydrograph is then used to estimate runoff from precipitation events of greater or lesser volume.

The SCS curve number and unit-hydrograph methods are lumped-parameter analytical approaches to runoff estimation. Lumped-parameter methods apply a single parameter value to the entire watershed area of interest.

Distributed-parameter rainfall-runoff numerical codes are more appropriate for larger watersheds with heterogeneous flow characteristics. HEC-HMS software, developed by the U.S. Army Corps of Engineers Hydrologic Engineering Center (Scharffenberg and Fleming 2010), simulates precipitation-runoff processes in dendritic watershed systems, ranging from small watersheds to large river basins. The Precipitation Runoff Modeling System (PRMS; Markstrom et al. 2008) was developed by the U.S. Geological Survey (USGS) to evaluate the streamflow response of a watershed to various combinations of climate and land use.

Surface water flow and transport models may be sufficient for prediction without incorporating reactive geochemistry into the model framework. Water quality in well-mixed rivers and streams can be predicted using a one-dimensional (1-D) channel-flow code such as QUAL2K (Chapra et al. 2008) or OTIS (One-Dimensional Transport with Inflow and Storage; Runkel 1998).

QUAL2K represents a modernized version of QUAL2E (Brown and Barnwell 1987). QUAL2K, distributed by the USEPA, runs within the Microsoft Excel spreadsheet environment. The program can simulate 1-D surface water flow, changes in water quality along the flow path, and chemical interactions with bed sediments.

OTIS is a code used to characterize the fate and transport of waterborne solutes in streams and rivers. The governing equation is the advection-dispersion equation with additional terms to account for transient storage, lateral inflow, first-order decay, and sorption.

For cases when reactive transport is important in surface water, OTEQ (Runkel 2010) couples OTIS with the chemical equilibrium code MINTEQ. Within OTEQ, reactions in the water column may result in the formation of solid phases (precipitates and sorbed species) that are subject to downstream transport and settling processes. Solid phases on the streambed may also interact with the water column through dissolution and sorption/desorption reactions.

More comprehensive, integrated watershed models may be developed that incorporate the surface-water elements of the hydrologic model into a broader framework. Codes used to implement integrated models at the watershed scale are described in more detail in the following section.

GROUNDWATER MODELS

The term *groundwater model* is widely used, and sometimes misunderstood. For the purposes of this document, *groundwater model* refers to a model of flow and nonreactive contaminant transport in a water-saturated zone beneath the ground surface. This encompasses conceptual groundwater models as well as physical analogues, analytical equations, analytic element approaches, or full numerical simulations. Models of water flow in unsaturated porous media are described in the "Vadose Zone Models" section later in this chapter. Reactive transport models are described in Chapter 7.

Advective Groundwater Flow

The basic equation of groundwater flow is referred to as Darcy's law (Darcy 1856). Numerous textbooks on groundwater hydrology describe Darcy's experiment and the resulting empirical equation (e.g., Fetter 1993; Freeze and Cherry 1979):

$$Q = -K \frac{dh}{dl} A \qquad \text{(EQ 6.4)}$$

where

Q = volumetric flow rate, volume/time

K = hydraulic conductivity, length/time

h = hydraulic head, expressed in units of length

l = distance between measuring points along the flow path, length

A = cross-sectional area perpendicular to flow

Dividing the volumetric flow rate Q by the cross-sectional area results in the specific discharge or Darcy flux, q:

$$q = -K\frac{dh}{dl} \quad \text{(EQ 6.5)}$$

For geochemical problems involving travel time, the pore velocity or average linear velocity v is important because it describes the velocity of water traveling along flow paths:

$$v = -\frac{K}{n_e}\frac{dh}{dl} \quad \text{(EQ 6.6)}$$

where n_e is the effective porosity (dimensionless fraction).

In a three-dimensional (3-D) Cartesian coordinate system, hydraulic conductivity is a second-rank tensor (Zheng and Bennett 2002):

$$\boldsymbol{K} = \begin{bmatrix} K_{xx} & K_{xy} & K_{xz} \\ K_{yx} & K_{yy} & K_{yz} \\ K_{zx} & K_{zy} & K_{zz} \end{bmatrix} \quad \text{(EQ 6.7)}$$

where K_{xx}, K_{yy}, and K_{zz} are principal components of the hydraulic conductivity tensor $\boldsymbol{K}$ and the remaining elements are cross-terms of $\boldsymbol{K}$.

In most practical applications, it is assumed that the principal components of the hydraulic conductivity tensor are aligned with the horizontal and vertical coordinate axes, reducing the cross-terms to zero (Zheng and Bennett 2002). Hydraulic conductivity is then represented by a vector with components K_x, K_y, and K_z that describe the hydraulic conductivity in each axis direction. For water of uniform density and viscosity, Darcy's law in three dimensions can then be written in differential form as

$$\begin{aligned} q_x &= -K_x\frac{\partial h}{\partial x} \\ q_y &= -K_y\frac{\partial h}{\partial y} \\ q_z &= -K_z\frac{\partial h}{\partial z} \end{aligned} \quad \text{(EQ 6.8)}$$

In an isotropic system, the magnitudes of K_x, K_y, and K_z are equal. In anisotropic groundwater systems, the magnitudes of K_x, K_y, and K_z are not equal. Many aquifers are characterized by a certain degree of vertical anisotropy due to the depositional environment and layering of the geological matrix.

By incorporating continuity and conservation of mass principles, the differential equation of groundwater flow takes the form of (Zheng and Bennett 2002)

$$\frac{\partial}{\partial x}\left(K_x \frac{\partial h}{\partial x}\right)+\frac{\partial}{\partial y}\left(K_y \frac{\partial h}{\partial y}\right)+\frac{\partial}{\partial z}\left(K_z \frac{\partial h}{\partial z}\right)+q_s = S_s \frac{\partial h}{\partial t} \quad \text{(EQ 6.9)}$$

where

q_s = fluid source–sink term, representing the volumetric rate at which water is added to or removed from the system per unit volume of aquifer

S_s = specific storage, representing the volume of water released from storage in a unit volume of aquifer per unit decline in head

t = time

Diffusion, Dispersion, and the Advection-Dispersion Equation

Molecular diffusion, the movement of solutes along a concentration gradient in the absence of advective flow, results in the spreading of a solute slug with time. Fick's first law describes the mass flux of solute per unit area due to molecular diffusion. Fick's second law describes systems where solute concentrations are changing with time.

At the scale of individual pores, groundwater flows at varying velocities because of differences in the individual flow paths. Solutes contained in the groundwater will also move at different velocities, causing mixing at the macroscopic scale of flow within an aquifer. This mixing is referred to as *mechanical dispersion* (Fetter 1993). Mechanical dispersion results in dilution of the solute at the advancing front of the flow system. Mixing along the direction of groundwater flow is referred to as *longitudinal dispersion*. Mixing perpendicular to the direction of groundwater flow is referred to as *transverse dispersion*. Coefficients of longitudinal and transverse dispersivity generally have different magnitudes.

The processes of molecular diffusion and mechanical dispersion cannot be separated in practical, field-scale groundwater problems (Fetter 1993). A combined parameter, the hydrodynamic dispersion coefficient, is used to represent the effects of both processes in most groundwater models. The hydrodynamic dispersion coefficient $\boldsymbol{D}$ in a 3-D system is represented by a second-order tensor (Zheng and Bennett 2002):

$$\boldsymbol{D} = \begin{bmatrix} D_{xx} & D_{xy} & D_{xz} \\ D_{yx} & D_{yy} & D_{yz} \\ D_{zx} & D_{zy} & D_{zz} \end{bmatrix} \quad \text{(EQ 6.10)}$$

Because the flow direction may not be aligned with the coordinate axes and the values of longitudinal and transverse dispersivity can be different, the values of the cross-terms of the hydrodynamic dispersion tensor cannot be assumed to be zero. Gelhar et al. (1992) provide a thorough review of field-scale dispersion in groundwater systems.

Solute transport due to advection, dispersion, and diffusion with internal sources or sinks is represented by the advection-dispersion equation (Zheng and Bennett 2002):

$$\frac{\partial(\theta C)}{\partial t} = \nabla \cdot (\theta D \nabla C) - \nabla \cdot (qC) + q_s C_s \quad \text{(EQ 6.11)}$$

where

$C =$ solute concentration in the aquifer

$C_s =$ solute concentration of a source or sink

$\theta =$ total porosity

$\boldsymbol{q} =$ Darcy velocity vector in three dimensions

$\nabla =$ grad operator in vector mathematics

$\nabla \cdot =$ divergence operator in vector mathematics

The advection-dispersion equation forms the basis for many groundwater flow and solute transport models over a wide variety of scales.

MODELING APPROACHES AND CODES

A large and ever-expanding body of literature exists regarding the topic of groundwater modeling. Javendel et al. (1984) describe analytical, semi-analytical, and numerical approaches to developing groundwater models along with a section on data needs and choice of methods. Wang and Anderson (1982) provide an introduction to the topic of numerical groundwater modeling, including derivation of the differential equations and the application of finite-difference and finite-element methods to problems of groundwater flow. Zheng and Bennett (2002) describe contaminant-transport modeling. Maest and Kuipers (2005) review groundwater models with an emphasis on prediction of acid drainage. Frippiat and Holeyman (2008) discuss upscaling methods for solute transport in heterogeneous porous media. Zhu and Anderson (2002) include a chapter on reactive transport modeling within the general framework of geochemical modeling.

Groundwater models should be as simple as possible while still representing the conceptual model with sufficient detail and accuracy. More complex models should only be developed and employed when required for a particular problem, assumptions inherent in simpler approaches are demonstrated to be violated, and/or appropriate data are available for model parameterization and calibration.

Ranging from simple to complex, three general categories of quantitative groundwater models can be applied to a given problem:

1. Analytical methods that use groundwater flow equations such as Darcy's law
2. Analytic element approaches
3. Full numerical methods

Each of these approaches to developing a groundwater model is described in the following sections, along with examples of computer codes that make use of these methodologies.

Analytical Methods

Analytical methods are closed-form equations that can be solved directly without the use of numerical methods. Transient or steady-state models of groundwater flow and contaminant transport in one, two, or three dimensions can be developed using analytical methods.

Models developed using analytical methods, such as the advection-dispersion equation, have relatively simple data requirements. Homogeneous, isotropic flow conditions are usually assumed. Analytical methods can be useful for screening-level evaluations and to determine the relative importance of different processes.

A useful analytical method for the prediction of impacts from acid drainage is the Ogata and Banks (1961) solution to the advection-dispersion equation. Domenico and Schwartz (1990) added a retardation factor to solve transport problems involving dissolved constituents that may react with the solid aquifer matrix according to a linear adsorption isotherm. The Ogata and Banks (1961) and Domenico and Schwartz (1990) equations can be solved in spreadsheets to address a wide variety of groundwater problems.

The STudio of ANalytical MODels (STANMOD; Šimůnek et al. 1999) is a publically available, free set of analytical solutions to the advection-dispersion equation in one, two, or three dimensions. A variety of previously published solutions is included in STANMOD.

Analytic Elements

Analytic element approaches use the principle of linear superposition to solve groundwater flow and contaminant transport problems in more complex systems than can be addressed by analytical equations. Haitjema (1995) provides the basic theoretical framework for the analytic element method and focuses on its use. Fitts (2010) describes a method of dividing an analytic element model into subdomains to allow representation of heterogeneous systems with spatially varying anisotropy.

GFLOW (Haitjema Software 2016) is a code that implements the analytic-element method for groundwater flow modeling. WhAEM2000 (USEPA 2007) is a public-domain and open-source analytic element code distributed by the USEPA. An AqSim (Fitts Geosolutions 2017) is an analytic element software package for simulating groundwater flow. VisualAEM (Craig 2009) is freely distributed software that provides a graphical user interface (GUI) for analytic element modeling of two-dimensional (2-D) and multilayer groundwater flow systems. It also allows numerical modeling of 2-D contaminant transport.

Numerical Methods

Numerical methods often use iterative processes to solve the equations of large-scale groundwater flow and contaminant transport in complex domains. Steady-state or transient flow in heterogeneous, anisotropic systems with varying types of boundary conditions and input stresses can be simulated. As model complexity increases, the amount of data regarding parameters and inputs also increases.

Finite-difference or finite-element solution schemes are used in the commonly available numerical groundwater flow codes. Wang and Anderson (1982) provide a thorough description of both solution schemes.

Finite-difference solution schemes can be more computationally efficient for a given problem compared to finite-element schemes. However, finite-element formulations can represent irregular external boundaries more precisely. They can also handle internal boundaries such as mine pits, underground workings, or faults with greater ease than finite-difference schemes. Because of the hydrogeological complexity of many mining-related flow systems, finite-element methods are often preferred.

The most commonly applied finite-difference groundwater modeling code is MODFLOW (Harbaugh 2005). Recent advances in the MODFLOW family of codes include MODFLOW-LGR, which allows local grid refinement, and MODFLOW-USG, which implements an unstructured grid method to support a wide variety of grid types. MODFLOW is available in the public domain from the USGS. It is probably the most widely used numerical groundwater code in the world.

Finite-element codes are exemplified by FEFLOW and MINEDW. FEFLOW (DHI-WASY 2012) is a commercially available code that can be applied to a broad range of variably saturated flow and transport problems. MINEDW (Itasca Denver Inc. 2012) was originally a proprietary code developed by Hydrologic Consultants Inc. for application to mining-related groundwater flow problems and is now distributed commercially by Itasca Denver Inc.

Many other finite-difference and finite-element models suitable for application to acid rock drainage (ARD)-prediction problems are available. The site conceptual model, project needs including regulatory or third-party review requirements, and available computing power are key elements to consider when selecting a solution method and computer code. Maest and Kuipers (2005) tabulate the capabilities of a range of codes.

Fracture Flow

The groundwater modeling approaches previously discussed are appropriate for simulating flow and transport through continuous porous media. However, the conceptual model for a particular problem may show that flow through discrete fractures or fracture networks is an important component. An equivalent porous medium may be defined in some of these cases if the fracture density is relatively high and the fracture aperture is small. Some codes also allow a dual-porosity formulation to consider flow through a fracture network as well as flow through the porous medium in blocks between fractures.

For cases where flow through individual fractures is important or a dominant mechanism, codes that represent discrete fracture networks and the physics of flow within fractures should be applied. An example of such a code is FracMan (Golder Associates Inc. 2012), which is commercially available. Peratta and Popov (2006) describe an approach to couple a discrete-fracture flow model with a conventional porous-medium model.

Data collection to properly parameterize a fracture-flow model is in most cases not included in routine hydrogeologic investigations. If the site conceptual model indicates fracture flow is significant, appropriate field procedures should be implemented.

Example Application

Wels et al. (2000) provide an example of groundwater model application to a site in New Mexico using MODFLOW in conjunction with other flow and geochemical models. In this case study, a model was implemented to estimate the dilution potential for tailings leachate in an underlying aquifer. A separate model using the variably saturated flow code SEEP/W was used to identify seepage pathways through the tailings and estimate seepage flux. MINTEQA2 was used to estimate long-term pore-water quality within the tailings due to sulfide oxidation and secondary mineral formation.

SURFACE-WATER/GROUNDWATER INTERACTION

Surface water commonly connects hydraulically to groundwater, but the interactions are difficult to observe and measure (Winter et al. 1998). Areas where surface water and groundwater interact provide opportunities for waters of differing geochemical composition to mix. For example, groundwater may discharge to a nearby surface stream. This could occur from diffuse discharges through the bed of the stream or point discharges at individual springs. Figure 6.2 shows the effects from influx of metal-rich groundwater from natural springs (foreground) to the surface water of Cement Creek in Colorado.

Conversely, surface water can discharge to groundwater. Hydrologic models of surface-water/groundwater interaction are important components of the geochemical models used to evaluate these types of systems.

Courtesy of Briant Kimball, U.S. Geological Survey

FIGURE 6.2 Metal-rich groundwater entering into Cement Creek, Colorado

Streams that are hydraulically connected to groundwater may be either losing or gaining. The water-surface elevation in a losing stream (Figure 6.3) is greater than the connected groundwater table, resulting in water moving from the stream into the aquifer. The water-surface elevation in a gaining stream (Figure 6.4) is lower than the connected groundwater table, resulting in water moving from the aquifer into the stream.

Streams may be losing in one reach and gaining in another. Seasonal variability in flow may result in a stream reach changing from losing to gaining over time. This can be an important factor in stream–aquifer systems in mountainous terrain, where a normally gaining stream can change to losing during the spring snowmelt.

The portion of the streambed and aquifer where the waters mix is termed the *hyporheic zone* (Figure 6.5). Because of this mixing, the geochemical composition of water in the hyporheic zone may be different from either the surface stream or groundwater (Winter et al. 1998). If a stream is separated from the underlying aquifer by an unsaturated zone, it is referred to as a *disconnected stream* (Figure 6.6). Water can move from a disconnected stream into the aquifer by unsaturated flow processes, possibly resulting in mounding within the saturated zone beneath the stream.

Groundwater codes such as MODFLOW and FEFLOW, among others, include the capability to model stream–aquifer interactions for hydraulically connected systems. The level of detail in the individual modules can vary depending on the conceptual model of the particular site.

GSFLOW is a public-domain, integrated code developed by the USGS (Markstrom et al. 2008). GSFLOW combines the surface-water code PRMS and the groundwater code MODFLOW. It was developed to simulate coupled groundwater/surface-water flow in one or more watersheds by coupling flow across the land surface, within subsurface saturated and unsaturated materials, and within streams and lakes.

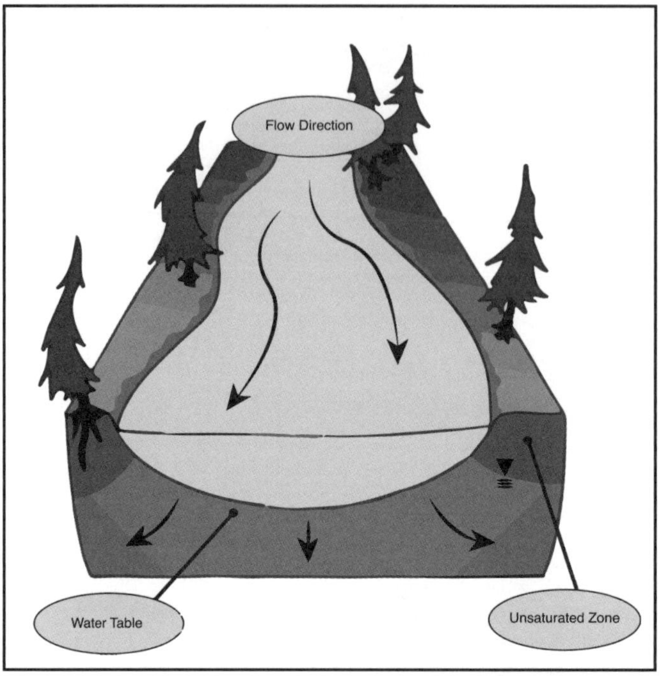

Adapted from Winter et al. 1998

FIGURE 6.3 Losing stream

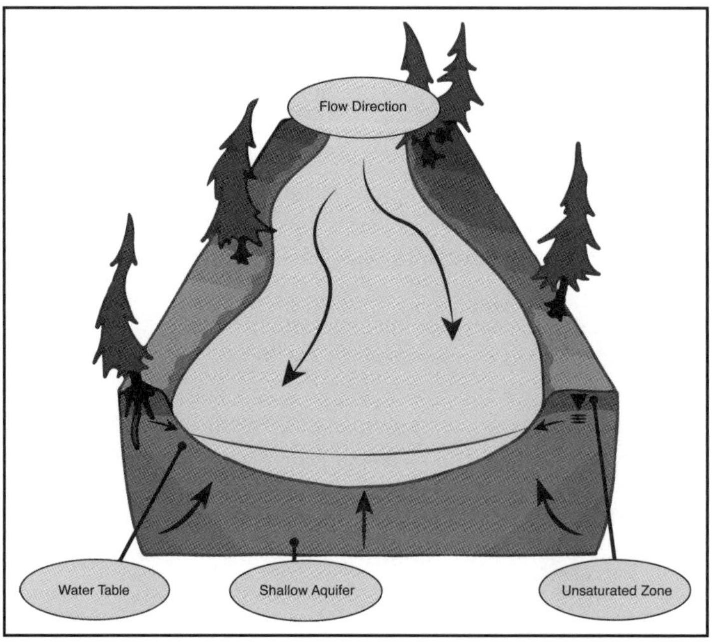

Adapted from Winter et al. 1998

FIGURE 6.4 Gaining stream

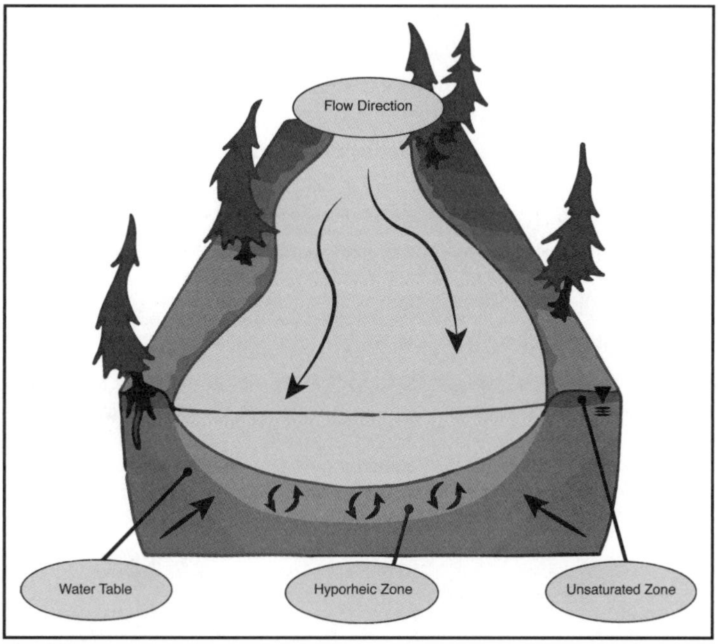

Adapted from Winter et al. 1998

FIGURE 6.5 Hyporheic zone

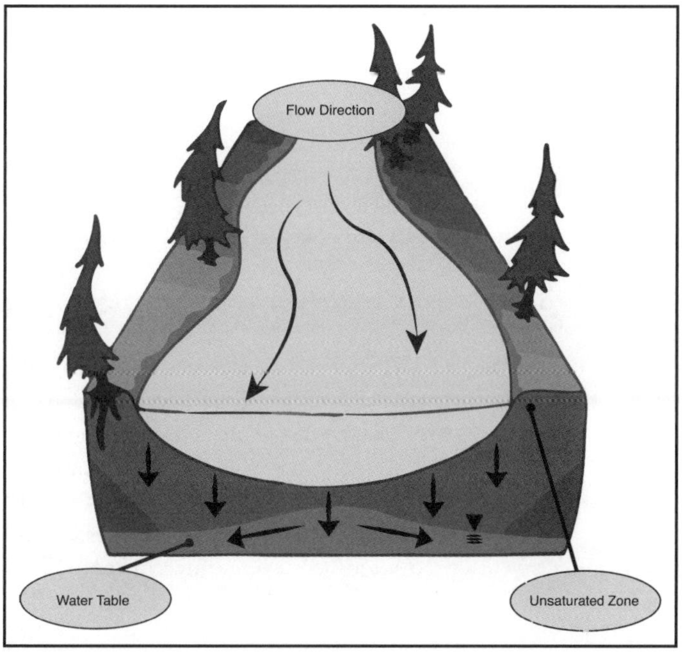

Adapted from Winter et al. 1998

FIGURE 6.6 Disconnected stream

Watershed codes, such as those described in the "Watershed-Scale Integrated Models" section later in this chapter, can be used to evaluate interactions between surface water and groundwater. For the case of a disconnected stream, a code capable of evaluating variably saturated flow within the vadose zone may be appropriate.

Interaction between lakes and groundwater can also be important, particularly in evaluations of pit lakes. Lakes can be sources (water moving from the lake to groundwater), sinks (water moving from groundwater to the lake), or flow-through elements. In a flow-through lake, water moves from the connected aquifer into the lake in one area and discharges from the lake into the aquifer in another area.

Vadose Zone Models

Potential impacts due to seepage through waste rock piles and tailings facilities are often evaluated using geochemical models. The flow components of these types of problems can be characterized by water flow through the vadose zone. The term *vadose zone* refers to the zone between the water table and land surface (Fredlund et al. 2012). This zone encompasses unsaturated porous media as well as the capillary fringe immediately above the water table.

Flow Equation

Darcy's law applies to the flow of water through an unsaturated porous medium (Richards 1931). However, hydraulic conductivity in unsaturated material is related to matric suction, which in turn is a function of moisture content for a given material (e.g., Richards 1931; Hillel 1980; Fredlund et al. 2012). Matric suction is a negative pressure potential attributable to capillary forces and the affinity of water for a particle surface due to effects such as adsorption (Hillel 1980). The equation for flow in unsaturated porous media can be generalized as (Hillel 1980)

$$q = -K(\psi)\nabla H \qquad \text{(EQ 6.12)}$$

where

ψ = matric suction

H = total hydraulic head, which may include both gravity and suction components

The relationship of matric suction to the degree of saturation is referred to as the soil–water characteristic curve (SWCC; e.g., Fredlund et al. 2012). The SWCC is a critical relationship to determine if unsaturated flow processes are important factors in the conceptual model for a given problem. Ideally, a site- and material-specific set of SWCCs will be determined in the laboratory. Well-defined procedures are available for this type of testing (Fredlund et al. 2012).

Determination of SWCCs can be time-consuming if multiple types of materials require characterization. In practice, SWCCs are often estimated from particle size distributions or by analogy to soils of similar texture. RETC (van Genuchten et al. 2009) is a computer program to estimate the hydraulic functions of unsaturated soils using different methods. SoilVision (Fredlund 2004) contains a database with unsaturated soils data for more than 6,200 samples.

Materials such as mine waste rock and tailings may have much different particle size distributions and structures than natural materials. Consequently, methods used in RETC or analogues in the SoilVision database should be applied with caution. Alternative approaches such as the physico-empirical model developed by Arya and Paris (1981) should be considered to account for the unique nature of mined materials. Aubertin et al. (1998) discuss predictive methods for SWCCs as applied to tailings from hard-rock mining operations.

Solution Methods

Because the governing equation is nonlinear, few analytical solutions exist for practical problems of unsaturated flow. Codes described in this section have been developed to implement numerical methods solving problems of unsaturated or variably saturated (combined unsaturated and saturated) flow. Some of these codes include GUIs and additional capabilities that are beyond the scope of this handbook.

HYDRUS (Sejna et al. 2011) is a finite-element code for simulating heat, flow, and solute transport in one, two, or three dimensions. The 1-D version of HYDRUS is available in the public domain. The 2-D and 3-D versions are available commercially.

SEEP/W (Geo-Slope International 2012a) is a finite-element software product for analyzing groundwater seepage and excess pore-water pressure dissipation problems within variably saturated porous materials such as soil and rock.

VADOSE/W (Geo-Slope International 2012b) is a finite element software product for analyzing flow from the environment, across the ground surface, through the unsaturated vadose zone, and into the local groundwater regime.

SVFlux (Fredlund 2012) is a commercially available code for simulating saturated and unsaturated flow in one, two, or three dimensions. It can be used to construct axisymmetric models in one or two dimensions and full finite-element models in three dimensions.

UNSAT-H (Fayer and Jones 1990) is a public-domain computer code used to simulate the 1-D flow of water, vapor, and heat in soils. It was developed at the Pacific Northwest National Laboratory.

MODFLOW-SURFACT (HGL 2012) is a commercially available, finite-difference code based on MODFLOW. It can be used to develop models of variably saturated flow and transport in one, two, or three dimensions.

Example Applications

Lefebvre et al. (2001a, 2001b) present an overview of numerical simulation methods applicable to waste rock piles. Fala et al. (2003) conducted a modeling evaluation of flow through idealized waste rock piles using HYDRUS to evaluate controlling mechanisms. Molson et al. (2005) used HYDRUS for the flow component of an acid drainage evaluation from waste rock piles.

PIT LAKES

Mine pit lakes represent the confluence of several elements of the water cycle. Groundwater, flow through the vadose zone, surface flow, direct precipitation, and evaporation may be of varying importance to a particular problem, depending on the physical setting of the pit lake. Vandenberg et al. (2011) present an overview of the use of models for the design and evaluation of pit lakes. Volume 3 of this handbook series (Castendyk and Eary 2009) presents a comprehensive review of mine pit lakes, including several chapters on hydrologic modeling of pit lakes.

The conceptual model for a pit lake is a critical element in guiding the development of appropriate mathematical models for the problem. Because of the complexity of pit lake systems, it can be tempting to develop overly complex mathematical models that suffer from a paucity of field data. However, a pit lake hydrologic model should start as simply as possible, with only the elements included that are shown to be important.

Many of the modeling approaches and codes previously described for water balances, surface water, groundwater, and the vadose zone can be and have been applied to pit lake problems. Watershed-scale models described in the following text may also be appropriate because they integrate most land aspects of the water cycle.

Because groundwater is often an important component of a pit lake water balance, specific approaches for these calculations have been developed. Marinelli and Niccoli (2000) summarize some of the analytical equations that are useful for estimating groundwater inflow to a pit lake. Aryafar et al. (2007) describe the development of a SEEP/W model to estimate mine pit inflows and compare the model to analytical solutions.

In addition to hydrologic models, limnologic models may also be developed for pit lakes. A limnologic model represents the physical and chemical processes that occur within a lake, such as mixing and stratification. Castendyk (2009) describes approaches to modeling the physical limnology of pit lake systems, including the application of the numerical code DYRESM (Dynamic Reservoir Simulation Model; Antenucci and Imerito 2001). CE-QUAL-W2 (Cole and Wells 2015) is a 2-D water quality and hydrodynamic code that has been applied to pit lake problems (e.g., Vandenberg et al. 2011). PITLAKQ (Mueller 2013) is an open-source code that couples CE-QUAL-W2 and PHREEQC.

WATERSHED-SCALE INTEGRATED MODELS

Continuing advances in computer processing power have allowed development of integrated models to simulate the land components of the full water cycle, including surface water, groundwater, infiltration, and evaporation, at the basin or watershed scale. Integrated models can be used to quantify the flux and volume components through different portions of the physical system within a single code framework. However, the data need to properly develop an integrated, quantitative model at this scale can be significant. Furman (2008) summarizes the mathematics and computational tools used to simulate coupled surface and subsurface flow processes. Examples of codes used to develop integrated models are described in the following text.

The Soil and Water Assessment Tool (SWAT) is a public-domain, distributed-parameter code developed by Texas A&M Agrilife Research and the Agricultural Research Service of the U.S. Department of Agriculture (Arnold et al. 2012). SWAT is designed to simulate runoff, groundwater flow, and water quality at the small watershed to river-basin scale. SWAT and similar codes break a complex watershed into hydrologic subunits, each with a uniform set of characteristics. Flow and water quality are calculated for each subunit, then aggregated to provide predictions at a complex watershed scale. Models using SWAT can also be developed within a geographic information system (GIS) framework.

MIKE SHE (DHI Group 2012) is a commercially available code that can be used to develop integrated models at the catchment scale. MIKE SHE can incorporate most processes in the land phase of the water cycle at varying levels of detail, depending on output needs. The MIKE SHE environment includes model pre- and post-processors along with some GIS functionality.

GAS PHASE

Gas transport mechanisms, including diffusion and advection, can transport oxygen into unsaturated waste-rock and tailings facilities. This can be an important factor in generation of acid drainage from these types of facilities. Wels et al. (2003) provide an overview of the role of gas transport in ARD generation and methods that can be used to model gas transport.

Diffusion in the gas phase can move oxygen (or other vapor-phase constituents) more effectively than diffusion in the aqueous phase. Free-air diffusivity coefficients are typically up to 4 orders of magnitude higher than the equivalent diffusivity coefficients in water. This difference in the magnitude of diffusive transport results in potentially creating conditions favorable for acid drainage in unsaturated media, while at the same time pointing to subaqueous disposal as a sometimes feasible option for limiting the generation of acid drainage.

Advective transport of oxygen into unsaturated media can occur because of convection as a result of temperature gradients and advection due to pressure gradients. Such pressure gradients can be caused by wind passing over a waste rock pile or because of differences between the ambient barometric pressure and the gas-phase pressure within the pore spaces of the unsaturated porous media. Massmann and Farrier (1992) describe the role of barometric pumping in the transport of gas into the vadose zone and demonstrate the capacity to transfer mass deep into an unsaturated porous medium. Amos et al. (2009) developed a conceptual model of advective transport due to wind-induced pressure gradients in a waste rock pile.

Gas-phase transport can be modeled using several of the codes previously mentioned for groundwater and vadose zone evaluations. Diffusion in the gas phase is included in unsaturated or variably saturated codes such as HYDRUS. Massmann (1989) describes how groundwater flow codes can be adapted to simulate flow in the gas phase. FEFLOW has the capability of modeling gas-phase flow.

Multiphase simulation programs have been developed that can be applied to problems involving gas flow. The TOUGH family of codes (Transport Of Unsaturated Groundwater and Heat; Pruess et al. 2012) was developed at Lawrence Berkeley National Laboratory to simulate the coupled transport of water, vapor, noncondensable gas, and heat in porous and fractured media. TOUGHREACT and TOUGH2-CHEM were applied to pyrite oxidation in variably saturated flow conditions by Xu et al. (2000). TOUGH-AMD (Lefebvre et al. 2001a, 2001b) is an adaptation of TOUGH to transport and reaction in waste rock piles.

CALIBRATION TOOLS

Model calibration is the process of selecting parameter values, inputs, and boundary conditions such that model output matches related observed data with an acceptable degree of accuracy and precision. Calibration of a hydrologic model is an application of inverse modeling. A significant body of literature exists regarding calibration of hydrologic models.

Calibration of a predictive model can be a significant effort. The level of calibration required for a particular model is dependent on the amount and type of data available and the precision or accuracy needed to meet project objectives. Hill and Tiedemann (2007) present suggested guidelines for effective model calibration along with a description of the calibration process. Vrugt et al. (2008) review the state of the science with respect to inverse modeling of subsurface flow and transport properties.

Several computer programs have been developed to automate the calibration process for particular codes. More recently, model-independent calibration or parameter-estimation programs have been developed. Two such programs that have been widely accepted for groundwater modeling are UCODE (Poeter et al. 2008) and PEST (Doherty 2010). Both UCODE and PEST have been incorporated into several GUIs to speed the model-calibration process.

Calibration of watershed models has historically been a manual process. More recently, advances have been made in automatic parameter-estimation methods applied to watershed models. Duan et al. (2003) summarize methods used to calibrate watershed models.

GRAPHICAL USER INTERFACES

A major development to simplify the application of computer codes to hydrologic models has been the advent of graphical user interfaces (GUIs). Several GUIs have been developed for public-domain programs, and most commercial codes include their own proprietary GUIs. Features such as GIS interfaces, parameter-estimation tools, and mesh generators are routinely included in GUIs for hydrologic codes.

GUIs provide significant capabilities, but can also be limiting. By insulating the user from the sometimes complex input files for a code, GUIs can greatly simplify and speed up the development of a model for a particular problem. They can also provide post-processing tools. However, GUIs can be limited in the number of modules supported for any particular computer code—for example, not all modules of MODFLOW are supported by every MODFLOW GUI. Some GUIs intercept and modify the underlying input files for the model code, making it difficult to troubleshoot a model and reducing the interoperability of models between different GUIs.

A selection of widely available GUIs is summarized in the following paragraphs. Others are no doubt available and may be suitable for hydrologic models used to help solve geochemical problems. Distribution of the modeling codes is typically included when purchasing a commercial GUI.

Groundwater Modeling System, Surface-water Modeling System, and Watershed Modeling System software products are commercially available GUIs distributed by Aquaveo LLC (2016). They provide GUIs for a wide range of hydrologic modeling codes.

ModelMUSE is a public-domain GUI for the MODFLOW family of codes. It was developed and is distributed by the USGS (2017).

Groundwater Vistas, developed by Environmental Simulations Inc., is a commercially available GUI for MODFLOW and related programs (ESI 2017). Visual MODFLOW Flex (Waterloo Hydrogeologic n.d.)is another commercially available GUI for the MODFLOW family of codes and related programs.

Selection of a GUI should be based on the capabilities of that particular GUI in relation to the type of problem being considered. Commercial and public-domain GUIs are under constant refinement to incorporate new capabilities and to keep up with the ongoing developments in hydrologic modeling practice.

CHAPTER 7

Reactive Transport Modeling

Ulrich Mayer

INTRODUCTION

Significant volumes of mine waste in the form of tailings or waste rock are generated during the mining process. Exposure of mine and processing wastes to precipitation and atmospheric oxygen induces rapid weathering reactions in particular because tailings and waste rock sometimes contain complex and highly reactive mineralogical assemblages, including sulfide minerals. In the absence of sufficient pH-buffering capacity, these weathering reactions produce low-pH waters containing elevated sulfate and metal loadings, typically referred to as acid rock drainage (ARD). Under neutral drainage conditions, high-sulfate loadings are also common, and select metals, including Cu, Zn, Pb, Mo, Se, and As, may be released at concentrations that pose environmental concern and require treatment.

Numerical models that are capable of linking the interactions between hydrologic transport and mineral weathering can be useful for assessing and predicting metal release from mine waste materials. Steefel and Van Cappellen (1998) have eloquently outlined the advantages of multicomponent reactive transport models as analytical tools for the investigation of natural systems. Among other important aspects, these authors point out the benefits of process-based reactive transport models over empirical models: the potential of these models to investigate coupled nonintuitive system behavior, the capability to conduct sensitivity analyses and "what-if" simulations, and to test conceptual models. More recently, Steefel et al. (2005) emphasized in their influential paper that "reactive transport modeling is an essential tool for the analysis of coupled physical, chemical, and biological processes in Earth systems"; the weathering of mine waste is a prime example.

The main purpose of this chapter is to introduce the basic concepts behind the reactive transport modeling approach. Many mine waste environments have in common that accelerated mineral weathering and metal release take place under unsaturated conditions, implying that the void space in the waste material is only partially filled with pore water, providing relatively easy access for oxygen from the atmosphere. Particular emphasis will therefore be given to formulations that are capable of simulating vadose zone water flow, gas and solute transport, coupled with geochemical reaction processes. Advantages and disadvantages of commonly used solution methods for reactive transport problems are also evaluated. Approaches for simulating metal leaching from mine waste are discussed; in particular, the author elaborates on setting realistic modeling goals and strategies for quantitatively evaluating conceptual models and assessing data requirements, and discusses limitations of process-oriented reactive transport models. Conceptual differences between modeling of metal release and attenuation in mine tailings and waste rock are also outlined. Furthermore, an illustrative example is presented to demonstrate the use of reactive transport modeling at a mine tailings site.

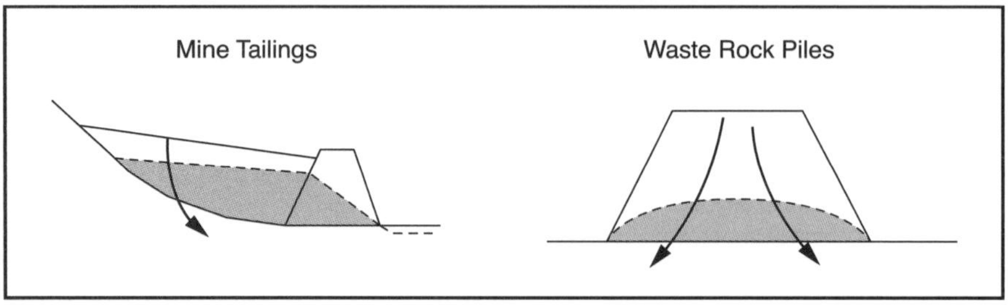

Source: Mayer et al. 2003

FIGURE 7.1 Mine waste environments that can be investigated using reactive transport modeling

GOVERNING PROCESSES IN MINE AND PROCESS WASTE ENVIRONMENTS

Mine and process waste environments can be subdivided into two main categories: mine tailings and waste rock piles (Figure 7.1). These environments have in common that weathering processes take place when these mine wastes are exposed to atmospheric conditions. The infiltration of dilute rain water or snowmelt causes the dissolution of mineral phases and at the same time facilitates metal transport with the infiltrating recharge (Figure 7.2). The weathering of mine waste is typically aggravated because of the presence of sulfide minerals that are commonly associated with ore deposits. Under unsaturated conditions, atmospheric oxygen is relatively easily replenished, causing a significant degree of sulfide mineral oxidation over short time frames. This process leads to the generation of acidity, as well as sulfate and metal release. These reaction products will accumulate in the pore water as it migrates downward; however, reaction products will also interact with non-sulfidic mineral phases present in the waste material, causing the dissolution of primary carbonates and silicates (Figure 7.2). In these highly reactive systems, the attenuation of metals within the waste material is common, including the formation of secondary minerals, co-precipitation, and adsorption. Understanding the combined effect of metal release and attenuation along the flow path is beneficial for assessing long-term metal loadings emanating from mine waste, assessing future environmental impact, and identifying appropriate operational and postclosure waste and water management alternatives, including treatment requirements.

Generalized Conceptual Model of Metal Leaching

As discussed in Chapter 2, prior to simulating the weathering of mine waste with a numerical model, it is a necessary requirement to conceptualize the problem at hand and to identify and classify the various reaction and transport mechanisms that are thought to be of importance at a site. Typically, evaluation of mine drainage issues requires the simulation of water flow under variably saturated conditions, solute transport, potentially gas transport, as well as a wide range of geochemical reaction processes that control metal release and attenuation. Figure 7.3 presents a generalized conceptualization of the reactive transport problem applicable to partially saturated mine waste where the water and gas phases are considered mobile and the solid phase is immobile (Mayer et al. 2002). Dissolved chemical species migrate with the mobile water phase, while atmospheric oxygen and carbon dioxide released from mineral weathering can move within or with the gas phase and partition in and out of the water phase. The solid phase is composed of minerals, amorphous phases, and surface species.

Geochemical reactions among these constituents may occur within a single phase (homogeneous reactions) or may involve two or more phases (heterogeneous reactions; e.g., Rubin 1983). Homogeneous reactions within the water phase (also known as the aqueous phase)

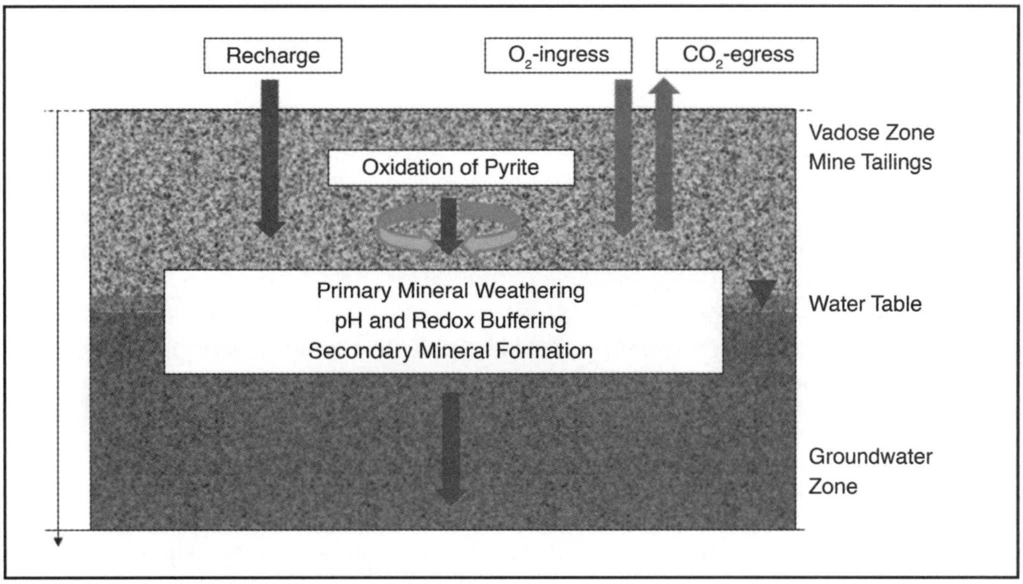

Source: Mayer et al. 2012

FIGURE 7.2 Conceptual model of sulfide mineral oxidation in mine tailings

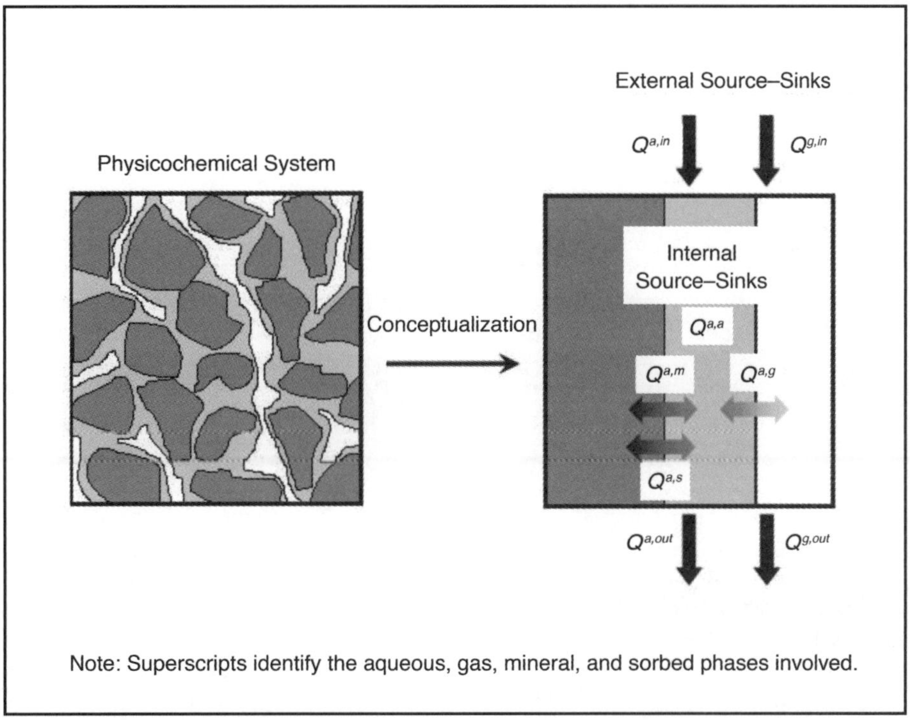

Adapted from Mayer et al. 2002

FIGURE 7.3 Conceptualization of physicochemical system

include hydrolysis, complexation, and aqueous oxidation–reduction reactions (e.g., oxidation of dissolved Fe(II) to Fe(III)). Heterogeneous reactions result in mass transfer between the phases and include mineral dissolution–precipitation (e.g., dissolution of carbonate and silicate minerals, sulfide mineral oxidation, and the formation of secondary hydroxides, oxides, and sulfates), ion-exchange, adsorption, and gas exchange between the pore water and the soil gas (e.g., partitioning of O_2 and CO_2 between the water and gas phases; Figure 7.3).

Geochemical reactions contribute to the addition or removal of solute to the pore water within a representative elementary volume (REV), which is mathematically described by internal source and sink terms. In Figure 7.3, these source–sink terms are defined as $Q^{a,a}$, $Q^{a,g}$, $Q^{a,m}$, and $Q^{a,s}$, where the superscripts identify the aqueous, gas, mineral, and sorbed phases involved. External sources and sinks ($Q^{a,in}$, $Q^{g,in}$, $Q^{a,out}$, and $Q^{g,out}$) are due to physical transport of the mobile constituents into or out of the REV and include advection, diffusion, and dispersion.

This general conceptualization captures the key processes controlling the geochemical evolution in mine tailings. It must be emphasized that waste rock poses additional modeling challenges due to its wide variation of grain sizes, high degree of physical and chemical heterogeneity, and complex structure caused by the mode of deposition. In this case, it may be necessary to consider the effects of preferential flow and/or internal heat generation on transport and reaction processes within the mine waste deposit.

Model Formulation for Multicomponent Reactive Transport

In this section, the key equations are introduced, describing water flow under variably saturated conditions, solute transport, gas transport, and geochemical reactions, closely related to the concepts introduced in Figures 7.2 and 7.3. In principle, by discretizing these equations in space and time, reactive transport models subdivide the domain of interest (e.g., a cross section through a tailings impoundment) into a series of connected REVs, as described in Figure 7.3, allowing simulation of geochemical changes within each cell and chemical mass fluxes between the cells.

For general background on groundwater flow and solute transport modeling, as well as reactive transport modeling in saturated systems, refer to Frind and Molson (1994). Further detail on the formulation of reactive transport equations can be found in the comprehensive reviews by Lichtner (1996) and Steefel and MacQuarrie (1996), or in Steefel et al. (2015).

Modeling variably saturated flow. Considering that metal release typically occurs in the unsaturated zone, it is beneficial to incorporate vadose zone flow processes into reactive transport models. In many circumstances, this can be achieved by using the Richards equation (e.g., Neuman 1973; Panday et al. 1993):

$$S_a S_s \frac{\partial h}{\partial t} + \phi \frac{\partial S_a}{\partial t} - \nabla \cdot \left[k_{ra} \boldsymbol{K} \nabla h \right] = Q_a \qquad \text{(EQ 7.1)}$$

where S_a is the saturation of the aqueous phase, S_s is the specific storage coefficient, h is the hydraulic head, t is time, ϕ is porosity, k_{ra} is the relative permeability of the water phase, and **K** defines the hydraulic conductivity tensor. Q_a is a source–sink term [s^{-1}], for example, for the description of infiltration. The relative permeability k_{ra} is often calculated using van Genuchten soil hydraulic function parameters, which provide a framework for describing soil–water characteristic curves (see Chapter 6 for more details) and provide the constitutional relationships between water pressure, saturation, and relative permeability (Wösten and van Genuchten 1988).

Equation 7.1 describes the movement of water under saturated and unsaturated conditions; the first term of the equation describes the change of water storage due the compressibility of water and the sediments, and is mostly relevant for saturated conditions. The second term describes changes in water storage due to the filling or draining of pore space and dominates under unsaturated conditions. Equation 7.1 provides a solution for hydraulic heads and allows the calculations of water saturations and Darcy fluxes required for solution of the transport problem.

Alternatively, a more sophisticated multiphase-flow formulation capable of considering the advective movement of the gas phase can be implemented (Lichtner 1996; Xu et al. 2000). Gas advection may result, for example, because of heat-induced pressure gradients (Lefebvre et al. 2001a, 2001b) and can provide an important transport mechanism in waste rock piles.

The formulation described by Equation 7.1 assumes uniform flow, implying that preferential flow due to physical material heterogeneity, which is often of importance for waste rock, is ignored. Various formulations, including dual porosity and dual permeability, are available in the literature (e.g., Šimůnek et al. 2003). However, to date, these formulations have not been invoked to simulate flow and reactive transport in mine waste.

Modeling reactive solute and gas transport. Solute transport is commonly described using the advection-dispersion equation. The use of the multicomponent-based approach allows the formulation of the transport problem in terms of total concentrations, rather than transporting each chemical species separately (Yeh and Tripathi 1989), which makes the problem computationally more tractable. The governing equation for the reactive transport of dissolved species in variably saturated media is

$$\frac{\partial}{\partial t}\left(S_a\phi T_j^a\right)+\frac{\partial}{\partial t}\left(T_j^s\right)-\nabla\cdot \boldsymbol{q}_a T_j^a+\nabla\cdot \boldsymbol{D}_a\nabla T_j^a=Q_j^{a,a}+Q_j^{a,m} \qquad j=1,N_c \qquad \text{(EQ 7.2)}$$

where T_j^a are the total aqueous component concentrations; T_j^s are the total sorbed component concentrations; $\mathbf{q}_a$ is the Darcy flux vector, which can be obtained from the solution of Equation 7.1; $\mathbf{D}_a$ is the hydrodynamic dispersion tensor applicable to all species in the system; and $Q_j^{a,a}$ and $Q_j^{a,m}$ are source–sink terms due to intra-aqueous and mineral dissolution–precipitation reactions. N_c defines the number of chemical components considered in the system (i.e., Fe(II), Fe(III), SO_4, H^+, etc.).

If the partitioning of gases between the water and gas phases is at equilibrium, Equation 7.2 can be extended to yield (Šimůnek and Suarez 1994)

$$\begin{aligned}\frac{\partial}{\partial t}\left(S_a\phi T_j^a\right)+\frac{\partial}{\partial t}\left(S_g\phi T_j^g\right)+\frac{\partial}{\partial t}\left(T_j^s\right)-\nabla\cdot \boldsymbol{q}_a T_j^a+\nabla\cdot \boldsymbol{D}_a\nabla T_j^a-\\ \nabla\cdot \boldsymbol{q}_g T_j^g+\nabla\cdot \boldsymbol{D}_g\nabla T_j^g=Q_j^{a,a}+Q_j^{a,m} \qquad j=1,N_c\end{aligned} \qquad \text{(EQ 7.3)}$$

where S_g is the saturation of the gas phase defined as $S_g = 1 - S_a$, T_j^g are the total component concentrations in the gas phase, $\mathbf{q}_g$ is the advective mass flux in the gas phase, and $\mathbf{D}_g$ is the dispersion tensor in the gas phase. In mine tailings, it is commonly possible to neglect advective gas transport; in this case, $\mathbf{D}_g$ simplifies to the diffusion coefficient for the gases.

Equation 7.3 maintains the mass balance for aqueous, gaseous, and sorbed species. In addition, it is necessary to provide a mass balance equation for the mineral phases:

$$\frac{d\phi_i}{dt}=V_i^m R_i^m \qquad i=1,N_m \qquad \text{(EQ 7.4)}$$

where ϕ_i is the volume fraction of the mineral in question, V_i^m is the molar volume, and R_i^m is the rate of mineral dissolution–precipitation; N_m is the number of minerals considered. The multicomponent reactive transport approach requires Equations 7.3 and 7.4 to be solved in a coupled manner.

At first sight, it is not obvious how geochemical reactions are considered in the transport equations. Essentially, the reactions terms are "hidden" in the total concentration and source–sink terms. The total concentration terms contain all reactions that can be described by equilibrium relationships, while source and sink terms are related to kinetic reaction processes, leading to the production or consumption of components.

Geochemical reactions. *Thermodynamic equilibrium reactions.* The total concentration terms in Equation 7.3 implicitly account for aqueous complexation, redox equilibria, ion exchange, surface complexation, and partitioning of gases between the water and gas phases. For example, the total component concentration for a component in the aqueous phase is defined as follows (Yeh and Tripathi 1989; Steefel and Lasaga 1994):

$$T_j^a = C_j^c + \sum_{i=1}^{N_{a,eq}} v_{ij}^{a,eq} c_i^{a,eq} \qquad j = 1, N_c \qquad \text{(EQ 7.5)}$$

where C_j^c are the concentrations of the components as species in solution, $N_{a,eq}$ defines the number of aqueous complexes, $\nu_{ij}^{a,eq}$ are the stoichiometric coefficients of the components in the complex, and $C_i^{a,eq}$ are the concentrations of aqueous complexes. Equation 7.5 provides the mass balance for all dissolved species of a specific component. For example, it accounts for all species containing Ca, including the free ion (Ca^{2+}) and any aqueous complex ($CaOH^+$, $CaCO_3$(aq), $CaSO_4$(aq), etc.). The concentrations of secondary species ($C_i^{a,eq}$) are determined based on the concentrations of the components as species in solution (Ca^{2+}, CO_3^{2-}, Fe^{2+}, etc.) by law of mass action relationships subject to the appropriate activity corrections:

$$C_i^{a,eq} = \left(K_i^{a,eq} \gamma_i^{a,eq}\right)^{-1} \prod_{j=1}^{N_c} \left(\gamma_j^c C_j^c\right)^{v_{ij}^{a,eq}} \qquad i = 1, N_{a,eq} \qquad \text{(EQ 7.6)}$$

where $K_i^{a,eq}$ is the equilibrium constant for the dissociation of the secondary aqueous species into components as species in solution, $\gamma_i^{a,eq}$ is the activity coefficient for the secondary aqueous species, and γ_j^c are the activity coefficients for the components as species in solution. Activity coefficients are commonly determined from the modified Debye–Hückel or Davies equation, or by the Pitzer equations for high-ionic-strength solutions (see the "Activity Coefficients" section of Chapter 3 for additional details). Similar relationships can be defined for the gas phase and the sorbed phase (for additional details, refer to Lichtner [1996] and Steefel and MacQuarrie [1996], or Mayer et al. [2002, 2003]).

Kinetic reactions. The source–sink terms in Equation 7.3 account for the loss or production of component mass due to kinetic processes. Although a thermodynamic formulation is more convenient in terms of model parameterization (information on parameters controlling reaction progress is not required), kinetic relationships must be chosen for reactions that tend to be slow in relation to the time scale of transport. These reactions often include mineral dissolution–precipitation and some oxidation–reduction reactions, in particular, microbially mediated processes. The most obvious example for the modeling of metal leaching from mine waste is the oxidation of sulfide minerals in the vadose zone. An equilibrium formulation is inadequate under these conditions, considering that sulfides cannot coexist with atmospheric oxygen under thermodynamic equilibrium. Enforcing equilibrium conditions would instantaneously deplete O_2 or the sulfide mineral phase. An example of a source–sink term for the gain or loss of aqueous component mass due to dissolution–precipitation reactions is given by

$$Q_j^{a,m} = -\phi \sum_{i=1}^{N_m} v_{ij}^m R_i^m \qquad j = 1, N_c \qquad \text{(EQ 7.7)}$$

where N_m is the number of mineral phases considered, v_{ij}^m define the stoichiometric coefficients of the components composing the mineral phase, and R_i^m is the rate of dissolution or precipitation of a mineral.

Unlike for equilibrium reactions, which can be defined uniquely based on the law of mass action relationships (Equation 7.6), rate expressions describing kinetic processes may take many forms. For example, for kinetic dissolution–precipitation reactions, it is useful to relate the reaction progress to the reactive surface area of the mineral. Dissolution–precipitation reactions may be controlled by the progress of the chemical reaction on the mineral surface or may be limited by the transport of reactants to the mineral surface (Figure 7.4). An example of such a transport-controlled reaction is the shrinking core model (Levenspiel 1998), which has been extensively used in mining applications (e.g., Davis and Ritchie 1986; Wunderly et al. 1996; Mayer et al. 2002). A symbolic rate expression for dissolution–precipitation reactions can be defined as

$$R_i^m = \left(\phi_i, S_i^m, D_i^m, k_i^m, K_i^m, C_j^c, C_k^{a,eq}, T_j^a\right) \qquad i = 1, N_m,\ j = 1, N_c,\ k = 1, N_{a,eq} \qquad \text{(EQ 7.8)}$$

where S_i^m is the reactive surface area of the mineral; D_i^m is a diffusion coefficient describing the migration of a reactant through a surface coating, required for the shrinking core model (e.g., Wunderly et al. 1996); k_i^m and K_i^m are the rate coefficients and solubility constants. This formulation allows, for example, the description of sulfide mineral oxidation or pH-dependent aluminosilicate weathering reactions. For a detailed description of implementing kinetic reactions in reactive transport models, refer to Mayer et al. (2002). The rate defined by Equation 7.8 will control change in mass in the mineral phase (see Equation 7.4) with a direct link to changes in mass in the aqueous phase (see Equations 7.6 and 7.3).

Solution methods for multicomponent reactive transport. In terms of coupling transport and reaction processes, two distinct formulations are commonly used for solving physical transport and geochemical reactions as described by Equation 7.3. These are the global implicit formulation or one-step approach, and the operator-splitting formulation, which also can be defined as the two-step or sequential approach (Engesgaard and Christensen 1988; Yeh and Tripathi 1989; Saaltink et al. 2001). The major difference between these methods is that physical transport and geochemical reactions are solved simultaneously using the global implicit method, while the operator-splitting technique considers transport and reactions sequentially with or without iteration between the two steps (Steefel and Lasaga 1994; Walter et al. 1994).

A solution of the global implicit method is commonly obtained using the direct substitution approach (Steefel and Lasaga 1994). In this case, the equilibrium expressions and/or kinetic rate equations are substituted directly into the transport equations as was done previously in Equation 7.3, yielding a single set of nonlinear partial differential equations, which are solved simultaneously using a Newton–Raphson-type linearization technique. This set of equations is

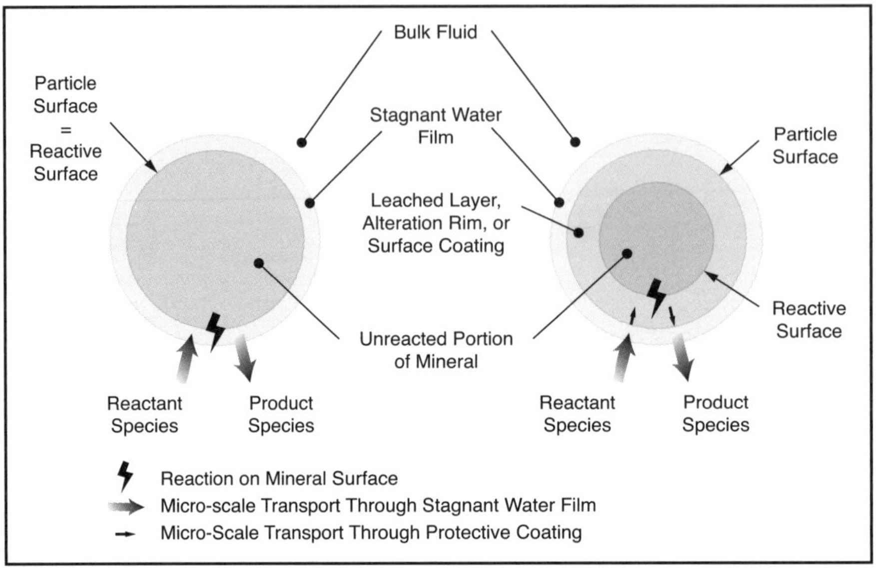

Source: Mayer et al. 2002

FIGURE 7.4 Surface-controlled reactions versus the shrinking core model

described by Equations 7.3 and 7.4 and was used by various authors, including Steefel and Lasaga (1994), Lichtner (1996), Xu et al. (2000), and Mayer et al. (2002).

Yeh and Tripathi (1989) present various methods for the solution of the two-step problem, which are based on the sequential iteration approach (SIA) or the sequential non-iterative approach (SNIA). This approach subdivides the reactive transport problem into subproblems, a physical and a chemical step (Walter et al. 1994; Frind and Molson 1994). The SNIA solves the transport Equation 7.3 by neglecting the source–sink terms, followed by the solution of geochemical speciation and kinetic processes in a separate step to obtain the concentrations at the new time level. The SIA uses a similar tactic but includes iteration between the two steps. This approach has also been used extensively (e.g., Walter et al. 1994). A detailed discussion of the formulation of the solution approaches is beyond the scope of this chapter. For a more comprehensive discussion on this topic, refer to Steefel and MacQuarrie (1996) and Saaltink et al. (2001).

For complex systems, the splitting of operators is often not restricted to the classical sequence of transport and geochemical reactions but might involve several steps yielding a multistep method as implemented in the MINTOX model (Wunderly et al. 1996). In this case, a three-step approach is used to solve ARD generation and attenuation. The first step consists of the simulation of gas transport and sulfide mineral oxidation and focuses on the quantification of contaminant release; the second step consists of the transport of all dissolved species, followed by the third step consisting of geochemical equilibrium calculations describing water–rock interactions. This method has also been adopted by Gerke et al. (1998).

Advantages and disadvantages of coupling methods. Several authors have discussed the advantages and disadvantages of the various solution approaches currently used in reactive transport modeling (Engesgaard and Christensen 1988; Yeh and Tripathi 1989; Mangold and Tsang 1991; Steefel and Lasaga 1994; Steefel and MacQuarrie 1996; Saaltink et al. 2001). Advantages of the two-step method are (based on the findings of Engesgaard and Christensen 1988, Steefel and Lasaga 1994, Walter et al. 1994, and Steefel and MacQuarrie 1996) a higher flexibility when dealing with a complex chemical system, which makes it easy to add or remove a process. In addition, this approach facilitates that individual routines for flow, transport, and reactions are exchangeable and it is not necessary to invert the large matrices generated by the simultaneous solution of a coupled problem. Generally the sequential approach is easier to program and the memory requirements are less than those for a fully coupled method. Also, the time discretization for the transport and the chemistry operator can be chosen independently (e.g., Prommer et al. 2003).

Advantages of the one-step method are the simultaneous treatment of all processes in time, physical space, and chemical reaction space. Also, the global convergence properties of the one-step method may be better than for the two-step method, and it is sometimes possible to take larger time steps compared to the two-step approach.

For saturated systems, it was traditionally accepted that the two-step method is computationally less expensive for a single time step and is usually the most effective method with respect to computer cost for a simulation (Yeh and Tripathi 1989). However, Saaltink et al. (2001) have demonstrated that this is not always the case, in particular if chemical species are strongly attenuated in comparison to the conservative transport velocity (i.e., if solid phases impose a strong control on the geochemical evolution, as is often the case in mine waste environments). In addition, reactive transport in mine waste deposits is often characterized by the rapid ingress of atmospheric O_2 through the gas phase, which is consumed by oxidation reactions, leading to a quasi-steady-state concentration profile. These conditions will also favor the one-step method (Mayer 1999).

MODELING APPROACH

Setting Goals—Sensitivity Analysis, History Matching, and Predictive Modeling

Process-based multicomponent reactive transport modeling is well suited to simulate the combined effect of metal leaching, solute transport, and natural neutralization and attenuation processes in unsaturated waste materials (e.g., Mayer et al. 2000, 2002; Romano et al. 2006). In addition, reactive transport modeling can be used to assess the fate of reaction products in the water-saturated portion of the mine waste deposit and in underlying or adjacent aquifers (Bain et al. 2001; Romano et al. 2006).

The approach taken to simulate a specific site will largely depend on the input data and time record of data available for comparison of field observations and model results. The objective of a modeling study can be subdivided into three broad categories: sensitivity analyses, history matching, and predictive modeling.

Sensitivity analyses can be used to reveal the actual controlling parameters and demonstrate the evolution of the geochemistry in mine waste and affected aquifers under a range of possible conditions, if the available field data are insufficient to provide a more definite analysis. An investigation by Glynn and Brown (1996) highlights many of the benefits and limitations of geochemical and reactive transport modeling at sites with limited spatial information.

If detailed site-specific information is available, it may be possible to develop an adequate site-specific conceptual model that will allow the matching of the historical geochemical evolution at the site. If successful, this approach provides increased confidence for the use of this model to predict the future hydrogeochemical evolution of the site. Chapter 2 of this handbook provides more information on setting realistic expectations for modeling. These aspects are of particular relevance to reactive transport modeling, given the relatively high level of complexity of this modeling technique.

Site-specific conceptual model. For any reactive transport modeling study, a site-specific conceptual model needs to be developed prior to starting a modeling analysis (see Chapter 2 for more detail on the importance of conceptual models). The conceptual model should be defined based on the current understanding of the processes active at the site and should utilize the available field data to allow any quantitative statement regarding the geochemical evolution of the pore water and weathering progress at the site. The questions that can be asked and answered with the modeling exercise will strongly depend on the comprehensiveness of this conceptual model and the supporting data.

When formulating the conceptual model, one should be aware that, at most sites, flow, transport, and reaction processes controlling the geochemical evolution are not fully understood or are simply too complex to be fully accounted for by a comparably simple concept. The conceptual model should therefore focus on the processes that are believed to be most important without oversimplifying the system to provide a representative approximation of reality. Logically, the modeling results must be viewed in the light of this approximation.

Data requirements. Because reactive transport modeling attempts to describe contaminant release and attenuation in mine waste environments on a process-oriented basis, detailed data are required to form the conceptual model and even more so to allow quantitative modeling. As in any other modeling exercise, it is necessary to define boundary and initial conditions, and to provide physical and chemical parameters that characterize the subsurface environment and control flow, transport, and reaction processes.

Ideally this implies that the following parameters are available:

- Initial condition: Moisture contents, mineralogy of the porous media downgradient of the infiltration zone, information on sorption capacity (e.g., cation exchange capacity), initial chemical composition of groundwater in this zone
- Boundary condition: Recharge rate and chemical composition of the infiltration water over time
- Physical parameters: Porosities, hydraulic conductivities, soil hydraulic function parameters, diffusion coefficients, and dispersivities
- Thermodynamic and reaction progress parameters: Equilibrium constants, selectivity coefficients, rate constants, and reactive mineral surface areas
- Secondary mineral phases forming due to the interaction of the infiltrating water with the minerals initially present in the porous medium

All of the parameters listed previously are important; however, the significance of mineralogical composition cannot be overemphasized (Bain et al. 2001; Zhu and Burden 2001; Zhu et al. 2002), as it ultimately determines the release and attenuation potential for ARD. Infiltration rates and moisture contents (i.e., a good understanding of the hydrogeologic system) are of equal importance, because these parameters will determine the migration rate of pore water and the ability of oxygen to enter the system through the gas phase.

Additional data are needed to compare the modeling results to the actual evolution at the field site. These data may include volume fractions of secondary minerals that have formed or depletion of primary mineral phases and spatial profiles of pore water and pore gas composition. The information may include microbial community structure and function. Only with these data will it be possible to test and calibrate a model and facilitate any type of predictive modeling.

Most of the time, not all of these data are available, and this must be taken into account when deciding which questions can be answered using reactive transport modeling.

Limitations of reactive transport modeling. Based on the previous discussion, it is clear that it is difficult to predict the long-term evolution of ARD and its attenuation. A reactive transport model is designed to provide a continuous description in space and time. It has to be kept in mind; however, that a numerical modeling study is based on the data obtained from the field.

Field data are usually characterized by a great degree of temporal and spatial sparsity, and therefore only provide snapshots in time and are limited to discrete locations in space.

It is also often difficult to quantify processes that will not impose a control on the evolution of the major ion chemistry, but they are nevertheless of importance. The mobility of trace metals, for example, can be very difficult to assess, because it may be affected by a large number of competing processes, such as co-precipitation, adsorption, and surface complexation. A distinction between these processes can be difficult in terms of data availability and model capabilities; the predictive capacity for trace metal transport and release is therefore diminished.

As discussed in more detail in Chapter 2, predictive modeling involves the extrapolation of a present condition to the future, which cannot currently be supported by data. Several authors have addressed the question of model validation (Konikow and Bredehoeft 1992; Oreskes et al. 1994; Narasimhan 1995). Oreskes et al. (1994) report that the validation of a model is only possible if the model can be described by a closed system. A field site is conceptualized in terms of a solution domain, which is subject to initial and boundary conditions. Imposing boundary conditions on the modeling domain converts the model from an open to a closed system. The validity of the model will then depend on the validity of the boundary conditions and the initial conditions.

Another important aspect affecting the defensibility of long-term predictions is the nonuniqueness of solutions to reaction transport problems. Nonuniqueness essentially means that two or more different model realizations can produce results, which fit observed data equally well. Reactive transport modeling is prone to this problem, because it depends on a large number of input parameters, which typically have a high degree of uncertainty. Variations in these input parameters might lead to comparable results for the period of investigation. For example, it often cannot be determined if a change in aqueous concentrations can be attributed to ion exchange reactions or is due to dissolution–precipitation reactions.

ILLUSTRATIVE EXAMPLE—MINERAL WEATHERING IN TAILINGS IMPOUNDMENTS

This case study demonstrates the use of reactive transport modeling to simulate the generation and attenuation of ARD for a tailings impoundment in a process-based fashion. The example focuses on the oxidation of sulfide minerals in an unsaturated mine tailings deposit (Mayer et al. 2012; see Figure 7.2 for schematic representation).

Sulfide minerals considered here are limited to pyrite. In addition, the tailings contain the primary mineral phases calcite, siderite, gibbsite, gypsum, K-feldspar, and biotite. Primary minerals correspond to the reactive minerals initially present in the mine tailings. Ferrihydrite, jarosite, and amorphous silica are included as potential secondary mineral phases. These secondary minerals are not initially present but are allowed to form if the solution becomes supersaturated with respect to these mineral phases. All mineral dissolution–precipitation reactions considered in this demonstrative example are presented in Table 7.1.

The geochemical reaction network involves 12 aqueous components (Al_3^+, Ca^{2+}, K^+, CO_3^{2-}, Cl^-, H_4SiO_4, H^+, $O_2(aq)$, SO_4^{2-}, HS^-, Fe^{2+}, Fe^{3+}). Equilibrium conditions are assumed for the iron and sulfur redox couples. The ingress of atmospheric oxygen into the vadose zone and the egress of CO_2 produced by carbonate mineral dissolution toward the ground surface are assumed to be dominated by diffusion. Aqueous complexation reactions (21 total) must be included to adequately capture mineral solubilities. The simulation is conducted for a 5-m (16-ft) deep one-dimensional (1-D) vertical domain through the tailings, discretized into 0.05-m control volumes. The maximum time step taken is 1 year, and the simulation is performed for a period of 10 years.

TABLE 7.1 Primary and secondary mineral reactions considered in acid rock drainage example

	Primary Minerals	log K_{sp}
Pyrite	$FeS_2 + \frac{7}{2}O_2(aq) + H_2O \rightarrow Fe^{2+} + 2SO_4^{2-} + 2H^+$	215.3
Calcite	$CaCO_3 \leftrightarrow Ca^{2+} + CO_3^{2-}$	–8.48
Siderite	$FeCO_3 \leftrightarrow Fe^{2+} + CO_3^{2-}$	–10.45
Gibbsite	$Al(OH)_3(am) + 3H^+ \leftrightarrow Al^{3+} + 3H_2O$	8.11
Gypsum	$CaSO_4 \cdot 2H_2O \leftrightarrow Ca^{2+} + SO_4^{2-} + 2H_2O$	–4.58
K-feldspar	$KAlSi_3O_8 + 4H^+ + 4H_2O \rightarrow K^+ + Al^{3+} + 3H_4SiO_4$	0.08
Biotite	$KAl_3Si_3O_{10}(OH)_2 + 10H^+ \leftrightarrow K^+ + 3Al^{3+} + 3H_4SiO_4$	12.99
	Secondary Minerals	
Ferrihydrite	$Fe(OH)_3 + 3H^+ \leftrightarrow Fe^{3+} + 3H_2O$	4.89
Jarosite	$KFe_3(SO_4)_2(OH)_6 + 6H^+ \leftrightarrow K^+ + 3Fe^{3+} + 2SO_4^{2-} + 6H_2O$	–9.21
Silica (amorphous)	$SiO_2(am) + 2H_2O \leftrightarrow H_4SiO_4$	–2.71

Source: Mayer et al. 2012

The porosity was set to 0.5 and the vertical saturated hydraulic conductivity is 1×10^{-6} ms^{-1}. Soil hydraulic function parameters (Wösten and van Genuchten 1988) are $S_{ra} = 0.05$ [–] (residual saturation), $\alpha = 3.5$ [m^{-1}], and $n = 1.4$ [–]. The free-phase diffusion coefficient for all dissolved species is 2.4×10^{-9} ms^{-2}, while 2.1×10^{-5} ms^{-2} is assumed to be representative for all gases. Tortuosity corrections and effective diffusion coefficients are calculated based on the Millington formulation in the aqueous and gas phases. A longitudinal dispersivity of 0.005 m is assumed.

A constant recharge rate of 9.5×10^{-9} ms^{-1} (corresponding to 300 mm yr^{-1}) is assumed with a constant head of $h = 2.5$ m at the base of the domain, resulting in a water table located at an intermediate depth within the domain. A mixed-type boundary condition is applied at the ground surface as the transport boundary condition. This boundary condition consists of a specified mass flux (third type) for all dissolved species and fixed concentration in an atmospheric boundary layer (first type) for O_2 and CO_2 in the gas phase. Recharge water is similar in composition to rainwater but contains elevated potassium accounting for fertilizer application often used at closed tailings facilities in support of revegetation efforts. Recharge water is in equilibrium with atmospheric O_2 and CO_2. Recharge and pore water chemical compositions initially present in the tailings are presented in Table 7.2. Pore water initially present in the tailings is assumed to be in equilibrium with amorphous silica, calcite, gibbsite, gypsum, and siderite. Initial volume fractions and effective rate coefficients for all primary minerals are provided in Table 7.3.

Initial mineral abundances are typical for mine waste with a relatively low sulfide content, limited presence of carbonate minerals, and elevated concentrations of silicate minerals. As a result, most mineral phases, with the exception of silicates, are only present in small quantities of much less than 1%. Notably, mineralogy and effective rate constants are loosely based on measured and/or calibrated parameters from a decommissioned tailings impoundment at Nickel Rim, Sudbury, Ontario (Mayer et al. 2002). Silicate mineral dissolution is strongly affected by kinetic limitations, reflecting the slow dissolution of these phases. All secondary phases and the primary phase gypsum are assumed to react rapidly, subject to quasi-equilibrium reactions. This approach is adequate if the rate of transport is slow in relation to the time scale of reactions. In this case, it is not necessary to accurately define the reaction rate coefficients. It is sufficient to use rate coefficients on the order of 1×10^{-7} mol L^{-1} s^{-1} or greater to approximate equilibrium behavior.

Results show water saturations range from fully saturated conditions in the lower half of the vertical profile to $S_a = 0.74$ near the tailings surface (not shown). Unsaturated conditions in the

TABLE 7.2 Boundary and initial conditions for acid rock drainage example*

Parameter	Recharge	Initial
Al^{3+}	1.3×10^{-8}	2.6×10^{-8}
Ca^{2+}	1.9×10^{-3}	1.4×10^{-2}
K^{+}	9.0×10^{-3}	9.0×10^{-3}
CO_3^{2-}	3.2×10^{-4}†	2.5×10^{-3}
Cl^{-}	1.1×10^{-4}	1.1×10^{-3}
H_4SiO_4	2.0×10^{-4}	1.9×10^{-3}
H^{+} (as pH)	5.0	7.0
O_2(aq)	0.21‡	−2.5§
SO_4^{2-}	7.5×10^{-3}	2.0×10^{-2}
HS^{-}	—	3.3×10^{-12}
Fe^{2+}	1.3×10^{-11}	1.5×10^{-4}
Fe^{3+}	5.4×10^{-5}	7.2×10^{-12}

Source: Mayer et al. 2012
*All concentrations are in units of [mol l^{-1}] unless otherwise noted.
†As pCO_2 [atm].
‡As pO_2 [atm].
§As pe.

TABLE 7.3 Primary and secondary mineral reactions considered in acid rock drainage example

Mineral	Volume Fraction [–]	k_{eff} [mol dm^{-3} s^{-1}]
Pyrite	1.37×10^{-2}	1×10^{-9}
Calcite	1.77×10^{-3}	5×10^{-8}
Siderite	3.49×10^{-4}	1×10^{-8}
Gibbsite	8.30×10^{-4}	1×10^{-8}
Gypsum	6.06×10^{-3}	At quasi-equilibrium
K-feldspar	2.68×10^{-2}	5×10^{-11}
Biotite	7.31×10^{-2}	1×10^{-11}

Source: Mayer et al. 2012

upper region of the tailings facilitate the ingress of atmospheric oxygen (Figure 7.5a). The oxidation of pyrite is restricted to the upper region of the tailings (Figure 7.5b), limited by oxygen availability (Figure 7.5a). Pyrite oxidation leads to the release of acidity and results in the dissolution of calcite. In turn, a series of mineral dissolution–precipitation reactions occur buffering pH. These phases include siderite, gibbsite, and jarosite (Figure 7.5b, c). The simulated pH-buffering sequence is consistent with the results of previous studies (Walter et al. 1994; Mayer et al. 2002). However, ferrihydrite, a mineral commonly observed in ARD-affected systems, does not precipitate. This is due to elevated potassium concentrations, favoring the precipitation of jarosite. Potassium stems from the assumed fertilizer application and is further supplemented from the dissolution of biotite and K-feldspar (not shown). Weathering of biotite and K-feldspar is enhanced locally due to the presence of low-pH conditions. The zone of active silcate dissolution is directly correlated to the section of the tailings that show formation of amorphous silica (Figure 7.5b).

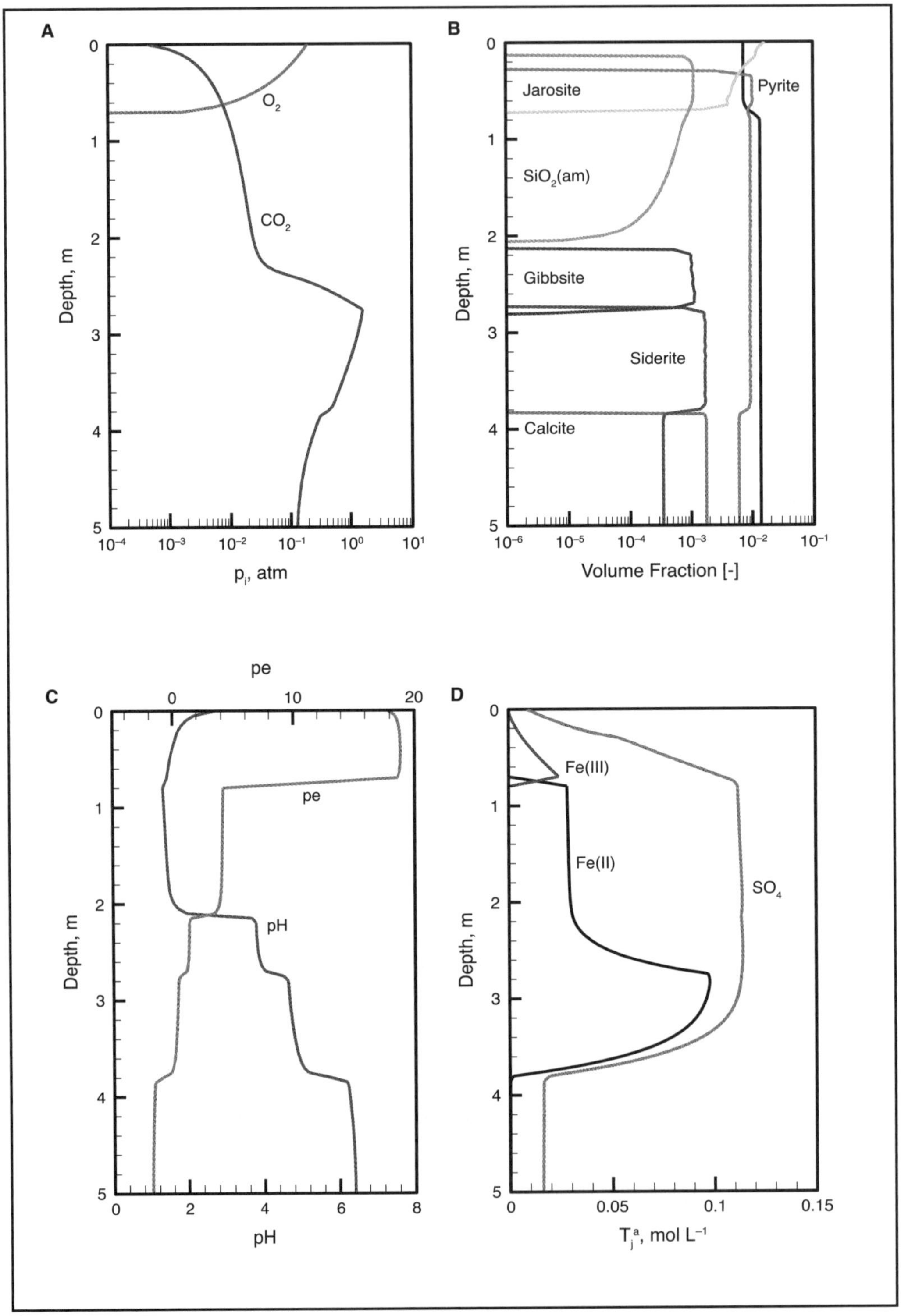

FIGURE 7.5 Oxidation of sulfide minerals in mine tailings. Simulation results as a function of depth; all profiles shown at T = 10 years: (a) O_2 and CO_2 partial pressures, (b) mineral volume fractions, (c) pH and redox potential, (d) total concentrations of Fe(II), Fe(III) and SO_6.

Carbonate mineral dissolution leads to increased CO_2 concentrations. If carbonate dissolution occurs above the water table, CO_2 tends to partition into the gas phase and is subseqently transported to the ground surface by gaseous diffusion. If carbonate dissolution occurs below the water table, CO_2 tends to remain in the water phase and migrates downward by advection (Figure 7.5a). The concentrations of iron and sulfate are strongly increased by pyrite oxidation, but jarosite and gypsum limit their solubility in pore water (Figure 7.5b, d).

A sharp transition in redox conditions can be observed where O_2 disappears. Ferric iron dominates in the region with O_2 present, while Fe(II) is dominant in the O_2-depleted region (Figure 7.5c, d). These steep gradients are due to the assumption of redox equilibrium.

This demonstration example is limited to a 1-D vertical profile; however, simulations are also possible for two-dimensional reactive transport problems (e.g., Romano et al. 2006).

SUMMARY AND CONCLUSIONS

This chapter provides an overview on the basic concepts for simulating reactive transport modeling in mine waste environments focusing on formulations applicable to variably saturated systems. Over the last 20 years, significant advances have been made in terms of incorporation of key processes into reactive transport models, including the incorporation of unsaturated flow or multiphase flow models (e.g., Acero et al. 2009), consideration of gas transport in the unsaturated zone, and comprehensive implementations of kinetically controlled reactions in addition to existing equilibrium formulations. Reactive transport models allow the simultaneous investigation of ARD and metal attenuation processes in various mine waste environments.

Increased model complexity has also led to increased data demand. It must always be kept in mind that modeling is only an approximation of reality and, therefore, it should not be used as a predictive tool if the supporting data are not available. The example presented here demonstrates that reactive transport modeling can be useful to describe and quantify key controls on the geochemical evolution, whether data availability is limited or a detailed data set is available. It is key to restrict the use of models to answering questions that can be supported by the available data. Possible modeling approaches include sensitivity analyses ("what-if" scenarios), history matching exercises to test conceptual models, and, with limitations, predictive modeling.

CHAPTER 8

Codes for Mine Site Characterization

D. Kirk Nordstrom

CODE LISTING

Computer software is constantly changing. Between advances in hardware and software, as well as growth in our understanding of natural and anthropogenic processes, programs will continue to get more sophisticated. The listing of codes in Table 8.1 is fairly current, but it is not intended to be comprehensive, and some references may be dated. To obtain the most recent information, the various codes should be searched online using the URLs provided or the code names.

Most of these programs are described in the recent review paper by Steefel et al. (2015). The reactive transport codes are described in the book by Zhang et al. (2012) along with the fundamental mathematics for fluid flow and chemical reactions for both equilibrium and kinetic conditions.

TESTING CODES

Testing codes is the first step toward getting some idea of code and model reliability. Unfortunately, it is not commonly done and it is rarely a regulatory requirement. Code developers demonstrate their codes with examples they have in their databank and they know what discrepancies to focus on. The code user typically does not test the code and does not know what to look for. Even the code developer cannot anticipate all possible conditions that might prove a computation incorrect. This is one of the better reasons why the codes and models they employ cannot be "validated" in any kind of meaningful sense. Codes can be tested and shown through a variety of techniques to provide the user with some confidence that the model is useful, but that is rather different from *validation*, that is, according to *Webster's New World Dictionary*, "having legal support or having proof and found free from error or criticism."

It is difficult to recommend a set of test cases for geochemical codes when the applications can vary widely for different site conditions and different objectives. Nevertheless, modelers and reviewers of models need to be aware that there are several published examples that can provide guidance as to what can be considered reasonable versus unreasonable computations. These examples are benchmarks or references against which new codes can be compared.

Seawater and River Water Test Cases

A seawater test case and a river water test case were designed to compare and contrast the speciation results from 11 geochemical or water chemistry codes (Nordstrom et al. 1979). Speciation, activity coefficients, and saturation indices of some common minerals were compared for both major ions and trace elements. For the river water test case, speciation results of major ions compared well, as might be expected for dilute water. Agreement was notably poor for the

TABLE 8.1 Computer codes used for mine waste problems

Code	Reference	URL
CE-QUAL-W2 (V 4.0)	Cole and Wells (2015)	www.cee.pdx.edu/w2/
CHESS	van der Lee et al. (2003)	http://chess.geosciences.ensmp.fr/
CrunchFlow	Steefel (2007)	www.csteefel.com/CrunchFlowIntroduction.html
EQ3/6	Wolery (1992); Wolery and Jarek (2003)	www.wipp.energy.gov/library/CRA/CRA-2014/References/Others/Wolery_Jarek_2003_Software_Users_Manual_EQ36_Version_8_0.pdf
The Geochemist's Workbench	Bethke and Yeakel (2014)	www.gwb.com/professional.php
GEOCHEMPC/ GEOCHEM-EZ	Shaff et al. (2010)	—
GoldSim	Hall et al. (2006)	www.goldsim.com/Web/Solutions/EnvironmentalSystems/
HYDROGEOCHEM	Cheng and Yeh (1998); Yeh et al. (2012)	www.scisoftware.com/products/hydrogeochem_overview/hydrogeochem_overview.html
HYDRUS, HP1/2/3, HPx	Šimůnek et al. (2013); Šimůnek et al. (2016)	www.pc-progress.com/en/Default.aspx?hydrus-3d
HYTEC	van der Lee et al. (2003)	http://pgt.geosciences.mines-paristech.fr/actualites/hytec-4-0-6-beta-disponible
LOADEST	Runkel et al. (2004)	http://water.usgs.gov/software/loadest/doc/
MIN3P	Mayer et al. (2002, 2003)	www.eos.ubc.ca/research/hydro/research/min3p/reactive_tran_web.htm
MINEWALL	MEND (1995); Perkins et al. (1995)	http://mend-nedem.org/wp-content/uploads/1152A.pdf
MINTEQA2	Allison et al. (1991)	www.allisongeoscience.com/MINTEQA2%20for%20Windows%20Document.pdf
NUFT	Hao et al. (2012)	https://wci.llnl.gov/codes/cafda/nuft_c/
OGS (OpenGeoSys)	Kolditz et al. (2012)	www.opengeosys.org/
ORCHESTRA	Van der Sloot and van Zomeren (2012)	www.macaulay.ac.uk/ORCHESTRA/
OTEQ	Runkel (2010)	http://water.usgs.gov/software/OTEQ/doc/
OTIS	Runkel (1998)	http://water.usgs.gov/software/OTIS/
PFLOTRAN	Hammond et al. (2012); Lichtner et al. (2015)	www.pflotran.org/docs/user_manual.pdf
PHAST	Parkhurst et al. (2010)	http://wwwbrr.cr.usgs.gov/projects/GWC_coupled
PHREEQC	Parkhurst and Appelo (2013)	http://wwwbrr.cr.usgs.gov/projects/GWC_coupled
PHREEQCi	Parkhurst and Appelo (2013)	http://wwwbrr.cr.usgs.gov/projects/GWC_coupled
PHT3D	Appelo and Rolle (2010)	www.pht3d.org/
PYROX	Wunderly et al. (1996)	www.flintbox.com/public/project/849
RATAP	Scharer et al. (1994)	—
RT3D	Clement and Johnson (2012)	http://bioprocess.pnnl.gov/rt3d.downloads.htm#doc
RETRASO, SA04	Saaltink et al. (2004)	http://h2ogeo.upc.edu/software/retraso/ManualRCB.pdf
SULFIDOX	Linklater et al. (2006)	www.asmr.us/Publications/Conference%20Proceedings/2006/1079-Linklater-AU.pdf
TOUGHREACT (V 3.0-OMP)	Xu et al 2004; Xu et al. 2012	http://esd1.lbl.gov/research/projects/tough/software/toughreact.html
WATEQ4F	Ball and Nordstrom (1991) with updates	http://wwwbrr.cr.usgs.gov/projects/GWC_chemtherm/software.htm

Courtesy of U.S. Geological Survey

trace elements. For the seawater test case, results were discrepant for some major ions and all trace elements. Discrepancies for trace element speciation were often orders of magnitude. The sources of these discrepancies, in order of importance, were the accuracy of the thermodynamic database, the number and type of complex stability constants, the method of calculating activity coefficients, redox assumptions, alkalinity corrections, and temperature corrections. Although thermodynamic data and activity coefficients have improved since then, they are still a major source of uncertainty for trace elements and redox species.

Speciation of surface seawater composition is a useful test case because not only is its composition constant, but it has been studied more carefully and intensively than any other water composition. In some studies, empirical measurements of activity coefficients and speciation were compared to theoretical calculations (see Millero 2001; Deng et al. 2011; Nordstrom 1996; Nordstrom and Campbell 2014). These comparisons provide a solid foundation for building confidence in our ability to calculate speciation. The seawater test case has been used as an example for such codes as WATEQ4F (Ball and Nordstrom 1991), PHREEQC (Parkhurst and Appelo 1999, 2013), EQ3/6 (Wolery 1992), and The Geochemist's Workbench (Bethke 2007, 2008). One of the most refined models for speciation of seawater composition has been incorporated into the spreadsheet speciation code from the University of Miami (Millero 2001).

Pyrite Oxidation

As explained in Chapter 5, pyrite oxidation is rather complicated because it involves (1) redox reactions for two elements, iron and sulfur, with a total transfer of 15 electrons; (2) hydrolysis of dissolved oxidized iron; (3) precipitation of a mixture of hydrated ferric oxides; and (4) differential kinetics for each reaction, of which some are biotic and some abiotic. The simulation of pyrite oxidation in Figure 5.1 makes a useful test case, in particular because there is some corroboration by the field measurements in Figure 5.2. Simulation of pyrite oxidation has been used as an example for the codes PHREEQC (Merkel and Planer-Friedrich 2008; Nordstrom and Campbell 2014), TOUGHREACT (Xu et al. 2000), and The Geochemist's Workbench (Bethke 2007, 2008).

Simulating pyrite oxidation with a computerized chemical model can be challenging, especially because site conditions will affect reaction rates. Hypothetical simulations like these help to understand the nature of the process, but attempting to apply these simulations to specific site conditions quantitatively usually requires more information than is available without a multitude of assumptions.

Other Examples in Code Manuals

Examples provided in code manuals serve two important purposes. They show how the code can be applied and they are useful in teaching new users how to set up input files.

PHREEQC and PHAST. The PHREEQC code has 18 examples of code application (Parkhurst and Appelo 2013). The most recent version is 3.3.11, released in 2017 (see Table 8.1). The examples are as follows:

1. Speciation of seawater
2. Solubility equilibrium transition (gypsum-anhydrite)
3. Mixing (seawater with carbonate groundwater)
4. Evaporation and homogeneous nitrogen redox reaction
5. Irreversible reactions (pyrite dissolution to equilibrium with calcite dissolution, release of CO_2 and goethite and gypsum precipitation)

6. Reaction path calculations (weathering of feldspar to gibbsite, kaolinite, and K-mica)
7. Gas-phase calculations (decomposition of organic matter)
8. Surface complexation (Zn sorption on hydrous ferric oxide)
9. Kinetic oxidation of dissolved ferrous iron with oxygen
10. Aragonite-strontianite solid solution
11. Transport and cation exchange (column flow with Ca-Na-K exchange)
12. Advective-diffusive flux of heat and solutes (column flow)
13. One-dimensional (1-D) transport with dual porosity and cation exchange
14. Advection with cation exchange, surface complexation, and mineral equilibria
15. 1-D transport with biodegradation, cell growth, and sorption
16. Inverse modeling: Sierra Nevada spring water
17. Inverse modeling: evaporation
18. Inverse modeling: Madison limestone aquifer

The PHAST code (Parkhurst et al. 2010) simulates multicomponent, three-dimensional, reactive solute transport for saturated groundwater systems. Seven examples from the literature, called verification examples, provided some confirmation that the code is working correctly. From these, four were chosen to provide more detail for the following tutorials:

1. Pulse source with sorption and decay
2. Chain of four kinetically decaying reactants
3. Aerobic consumption with microbial growth
4. Simulation of reaction and transport in regional scale aquifer (Central Oklahoma)

About half of the textbook *Groundwater Geochemistry* (Merkel and Planer-Friedrich 2008) is devoted to the theoretical background and available codes, and the other half to exercises and applications using the PHREEQC and PHAST codes. Several of the applications are as follows:

1. Seawater speciation
2. Gypsum solubility
3. Calcite solubility with CO_2 and open and closed systems
4. Pyrite oxidation with and without calcite
5. Mineral precipitation in hot springs
6. Carbonate stalactites in karst caves
7. Evaporation
8. Feldspar weathering
9. Organic matter degradation and reduction of inorganic redox elements
10. On-site leaching
11. Reactive transport modeling

EQ3/6. The code EQ3/6, in reality, consists of two codes: EQ3NR and EQ6. The code EQ3NR (Wolery 1992) computes speciation and equilibrium solubility. It has several examples that provide a tutorial as well as confirmation that the code is working correctly:

1. Speciation of seawater (with and without Pitzer model)
2. Calcite-hematite solubility constraints

3. Calculate pH of a buffer
4. O_2 fugacity for hematite-magnetite buffer at 250°C (482°F)
5. Eh computed from Fe redox couple
6. Dead Sea brine (water activity and solute activities with Pitzer model)
7. pH scales with Pitzer model

EQ6 (Wolery and Daveler 1992) calculates reaction paths (multiple reactions in a given process) with thermodynamic and/or kinetic constraints. Examples provided for EQ6 are as follows:

1. Computing multiple solubility equilibrium for seawater
2. Calculating high-temperature pH from quench pH
3. Microcline dissolution in pH 4 HCl
4. Microcline dissolution in open-flow system
5. Gypsum solubility in NaCl solutions (Pitzer model)
6. Alkalinity titration
7. Kinetics of quartz precipitation

TOUGHREACT. The code TOUGHREACT (Xu et al. 2006) is part of a series of code developments that combined geochemical reaction with a previously developed transport code. The following eight sample applications are provided in the user's guide (Xu et al. 2004):

1. Transport with reaction and decay
2. Geochemical evolution of groundwater in the Aquia aquifer (Maryland)
3. Yucca Mountain infiltration with calcite deposition (unsaturated zone, Nevada)
4. Thermal test at Yucca Mountain
5. CO_2 disposal in deep saline aquifers
6. Supergene copper enrichment
7. Hydrothermal alteration
8. Geothermal well scaling on injection (Philippines)

The Geochemist's Workbench. This code has an extensive user's guide (Bethke 2007) and an associated well-written textbook (Bethke 2008) that guides the reader through basic theory, reaction processes, and several applications, which include the following:

1. Hydrothermal mineralization
2. Geothermometry
3. Evaporation
4. Diagenesis
5. Water–rock kinetics
6. Weathering
7. Uranium reduction
8. Microbial degradation of phenol
9. Microbial reduction of arsenic
10. Autocatalytic manganese oxidation
11. Waste injection

12. Scaling and flooding in petroleum reservoirs
13. Acid mine drainage
14. Lead contamination
15. Groundwater chromatography and zoning

Benchmarks

Hydrogeochemical modelers have recognized the need to have well-described problems or benchmarks that allow rigorous testing of codes and models. A group of papers in a special issue of *Computational Geosciences* (Steefel et al. 2015) provides useful overviews and insights. A requirement for the papers in the special issue was that there had to be a comparison of at least three different codes per benchmark. Of course, this requirement does not imply that the models are different, only the codes. Nevertheless, they represent useful comparisons for examining how similar the results are for different codes applied to the same problem. The collection of special issue papers includes an introductory paper that lays out the fundamental mathematical equations of chemical reaction; microbial growth rates; physical flow with advection, diffusion, dispersion, single and multiphase flow, and the numerical formulation methods for solving the set of nonlinear equations. The remainder of the paper reviews 12 reactive transport modeling problems covered in the special issue, including a problem focusing on acid rock drainage (ARD) generation and attenuation.

The benchmark papers address the following:

1. Microbially mediated chromium reduction
2. Chromium isotope fractionation
3. Multicomponent diffusion and electrochemical migration
4. Simulations of column experiments
5. Uranium bioremediation
6. Biogenic uraninite reoxidation
7. U(VI) transport with surface complexation
8. Modeling of the generation and attenuation of ARD
9. Heavy metal cycling in lake sediments
10. Reactive transport across a cement–clay interface
11. Evaluation of permeability-porosity-tortuosity relationships with mineral dissolution and precipitation
12. Decalcification of cracked cement structures

The key conclusion of the special issue is that given the same database and treatment of activity coefficients, essentially identical results can be obtained with a variety of reactive transport codes, independent of their formulation details (Steefel et al. 2015). This should not be misinterpreted as predictive capability. This capability will ultimately be controlled by the validity of the site conceptual model and the quality and quantity of the input data. Nevertheless, the benchmarking exercises have built confidence in the numerical formulation of the participating codes.

DATABASES

The different levels of thermodynamic data are primary, secondary, and tertiary. Primary data are original papers of various thermochemical measurements such as calorimetry, solubility, electrochemistry, colligative properties, and phase equilibria. Secondary data are compilations of primary data. Tertiary data are compilations of secondary data (i.e., compilations of sources of thermodynamic data). The problem with compilations like secondary and tertiary sources is that there is often little or no evaluation of the original data. Only thorough evaluations provide useful references for databases. Selected sources of evaluated data are provided in Table 8.2.

Most geochemical modelers simply use the database that comes with a code; however, some codes such as PHREEQC, EQ3/6, and TOUGHREACT allow for a choice of databases in the same code. Having more than one database in a code puts more responsibility on the user to justify his or her selection of any particular database.

TABLE 8.2 Selected sources of evaluated thermodynamic data

Popular Name	Reference	Notes on Contents
NBS* tables	Wagman et al. (1982)	Comprehensive; no references (confer with National Institute of Standards and Technology)
USGS† tables	Robie and Hemingway (1995)	Minerals and aqueous species; comprehensive
NIST-JANAF‡ tables	Chase (1998)	Combustion reactants and products; critically assessed and referenced
NAGRA§ tables	Hummel et al. (2002); Thoenen et al. (2007)	Nuclear waste and groundwater components
OECD/NEA/TDB** tables	Book series††	U, Am, Tc, Np, Pu, Ni, Se, Zr, Fe, organics, and some groundwater components
CODATA‡‡ key values	Cox et al. (1989)	Elements, simple aqueous species, and a few related compounds; critically assessed
CODATA‡‡ Ca values	Garvin et al. (1987)	Compounds of Ca and related species; critically assessed
Critical stability constants	Martell and Smith (1974–1989)	Solubility and aqueous stability constants including organics
HKF§§ tables	Oelkers et al. (1995); Shock et al. (1997)	Minerals and aqueous species (similar to SUPCRT); critically assessed
SUPCRT***	Johnson et al. (1992)	Minerals and aqueous species
Geopig	Online database	http://geopig3.la.asu.edu:8080/GEOPIG/reactants.jsp; based on SUPCRT
Holland and Powell	Holland and Powell (2011)	High-temperature, high-pressure minerals and some aqueous species; critically assessed
IUPAC††† speciation series	Powell et al. (2005, 2007, 2009, 2011, 2013)	25°C (77°F) only; stability constants for Hg, Cu, Pb, Cd, and Zn with Cl^-, OH^-, CO_3^{2-}, SO_4^{2-}, and PO_4^{3-}; critically assessed

Courtesy of U.S. Geological Survey

*NBS = National Bureau of Standards (now National Institute of Standards and Technology)

†USGS = U.S. Geological Survey

‡NIST-JANAF = National Institute of Standards and Technology–Joint Army, Navy, Air Force

§NAGRA = Swiss National Cooperative for the Disposal of Radioactive Waste

**OECD/NEA/TDB = Organisation for Economic Co-operation and Development/Nuclear Energy Agency/ Thermodynamic Data Base

††Brown et al. 2005; Bruno et al. 2007; Gamsjäger et al. 2005, 2012; Grenthe et al. 1992; Guillamont et al. 2003; Hummel et al. 2005; Lemire et al. 2001, 2013; Olin et al. 2005; Rand et al. 2009; Rard et al. 1999; Silva et al. 1995

‡‡CODATA = Committee on Data for Science and Technology of the International Council for Science

§§HKF = Helgeson, Kirkham, and Flowers formulation (see Helgeson et al. 1981)

***SUPCRT92 = Supercritical 1992

†††IUPAC = International Union of Pure and Applied Chemistry

It has become increasingly important for modelers to be familiar with differences between databases and to document and justify their use of a particular database or particular equilibrium constants for any investigation. Sometimes a change in one or more equilibrium constants can change the interpretation and remedial decision about a site (Whiting 1992).

Table 8.2 summarizes some of the more commonly used thermodynamic databases and their references. Note that databases are dynamic and change as better data become available. However, newer databases are not necessarily more reliable. Also, some databases include mineral-phase properties only, others include minerals and aqueous species but not reactions, and yet others include reactions but not individual species. More details on databases can be found in the appropriate references and in the review of low-temperature geochemical modeling by Nordstrom and Campbell (2014).

Some sources of data have only aqueous species or reactions (Powell et al. 2005, 2007, 2009, 2011), some have only solid phases (Holland and Powell 2011), and some have both. Some sources compile data only for one element (Garvin et al. 1987; OECD/NEA/TDB book series listed in Table 8.2) whereas others focus on combustion products (Chase 1998).

A few words of caution. When looking up data, one must be familiar with thermodynamics and the methods of reducing raw measurements to thermodynamic properties. The many terms, symbols, and abbreviations used in thermodynamics are not necessarily consistent among different authors or institutions. Reactions are not always written the same way in different sources and it is, therefore, critical to have the correct reaction equation. Also, every measurement has sources of error and uncertainties. Analytical or precision errors do not necessarily reflect errors in accuracy. Lastly, new measurements do not necessarily supersede old measurements. Some of the most careful measurements on solubility were made toward the end of the 19th century and early in the 20th century.

In addition to sources of thermodynamic data, there are numerous compilations of unevaluated data. Many of these are listed in Nordstrom and Munoz (1994) up to 1993. This is likely to be the most extensive set of references available for that time period. Zemaitis et al. (1986) also provide an annotated list of references for useful data for aqueous electrolytes. Gradually over time, a convergence of data should evolve as both measurements and evaluations improve. Note, however, that even when multiple references present the same value for a thermodynamic property, it usually means that there is only one data source, and the accuracy of that value is not necessarily satisfactory.

The ultimate goal is to obtain internal consistency between all interrelated thermodynamic measurements (solubility, electrochemistry, calorimetry, isopiestics, phase equilibria, etc.). A geochemical goal would be to determine the range of conditions for which equilibrium could reasonably be applied and those for which they should not be. These are subjects of ongoing research.

CHAPTER 9

Case Studies

Questa Baseline and Premining Groundwater Quality

D. Kirk Nordstrom

The Questa molybdenum mine, located in northern New Mexico, has been producing molybdenum ore since 1923 and closed down in 2014. The mine consisted of extensive underground workings and an open pit that was mined from 1965 to 1983. As part of the permit requirements for closure, state law requires that the groundwater quality meet New Mexico standards unless it can be shown that the groundwater quality contained higher concentrations of contaminants naturally, before mining. At the request of the State of New Mexico, the U.S. Geological Survey completed a baseline and premining groundwater investigation to avoid any possibility of a conflict of interest with the mining company. The six-year scientific study was summarized in a final report (Nordstrom 2008) that highlighted results from 25 individual reports on the geology, hydrology, geophysics, and geochemistry of the Red River Valley where the mine is located.

Application

- Modeled saturation indices of key minerals to identify upper concentration limits of contaminants.
- Modeled mass flux of sulfate for groundwater to river and with transport in river.
- Solubilities of contaminants are very sensitive to pH.

Codes

- WATEQ4F
- BALANCE
- OTEQ
- PHREEQC

Processes

- Carbonate mineral dissolution and saturation
- Pyrite oxidation
- Gypsum dissolution/saturation
- Iron oxidation/reduction
- Ferrihydrite and jarosite precipitation
- Fluorite dissolution/saturation
- Alumina and silica saturation/precipitation
- Groundwater-surface water mixing
- Surface water reactive transport

Approach

- Used an unmined proximal analogue field site as a model.
- Correlated solutes and sulfate for acid (pH <4) groundwaters.
- Used saturation indices to find upper concentration limits of some constituents.

Lessons Learned

- Maximizing high-quality field data minimizes the need for poorly constrained computational modeling.
- Independent third-party studies can be advantageous to both regulatory and industry needs.

NATURAL ANALOGUE AS THE KEY

A catchment to the east of the mine property, known as Straight Creek (Figure 9.1), served as a proximal analogue field site for detailed hydrologic, geologic, and geochemical

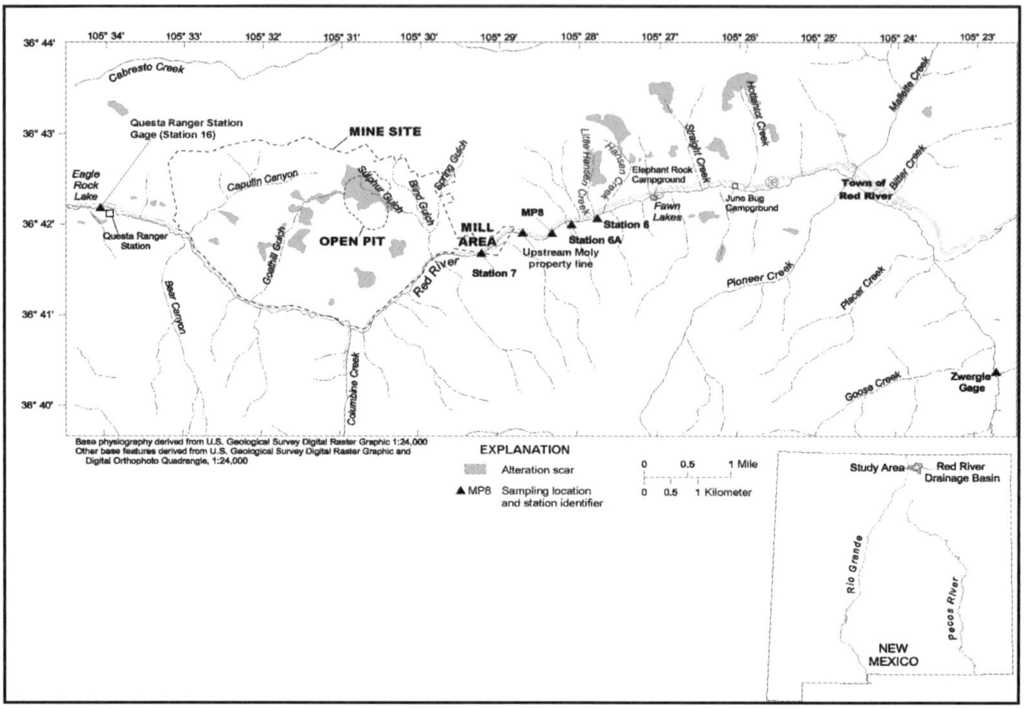

Courtesy of U.S. Geological Survey

FIGURE 9.1 Map of Red River Valley in northern New Mexico (see insert) with major features labeled. The natural analogue site, Straight Creek, is located just east of longitude 105° 27'. Scar areas are shown in gray.

studies. The catchment had very similar geological and hydrological characteristics as the mine site and was located only a few kilometers distant on the same south-facing slope of the Red River Valley. No mining had ever taken place here and there was an access road that made it possible to bring a drilling rig into the site. Altered volcanics (rhyolite and andesite) contain a substantial range of hydrothermal alteration from propylitic to quartz-sericite-pyrite (QSP). Acid rock drainage (ARD) occurs naturally from areas with QSP lithology. Because of rapid rates of erosion, large areas of mountain slope containing QSP are devoid of vegetation (scars) and large debris fans formed downgradient from them. In the debris fans, the groundwater was acidic (pH 3–4) and metal concentrations were elevated. Straight Creek was typical of these QSP lithologies with a debris fan that became about 150 m (492 ft) thick near the base at the Red River. Underneath the debris fans, volcanics contained abundant carbonates and were anoxic so that groundwater was circumneutral in pH and had low concentrations of metals, except for Fe, Mn, Ca, and Mg. Sulfate concentrations were also high and alkalinity was occasionally high. These waters often reached saturation with respect to gypsum, and secondary gypsum crystals could be frequently found in the bedrock and in debris material.

GEOCHEMICAL CONTROLS ON NATURAL ANALOGUE CONTAMINANT CONCENTRATIONS

In the Straight Creek acidic groundwaters, most constituents of concern, such as Zn, Cd, Mn, F, Be, Li, Co, and Ni, were shown to be conservative in acidic (pH <4) environments by their strong correlation with sulfate concentrations, assuming sulfate to be conservative (a safe assumption at

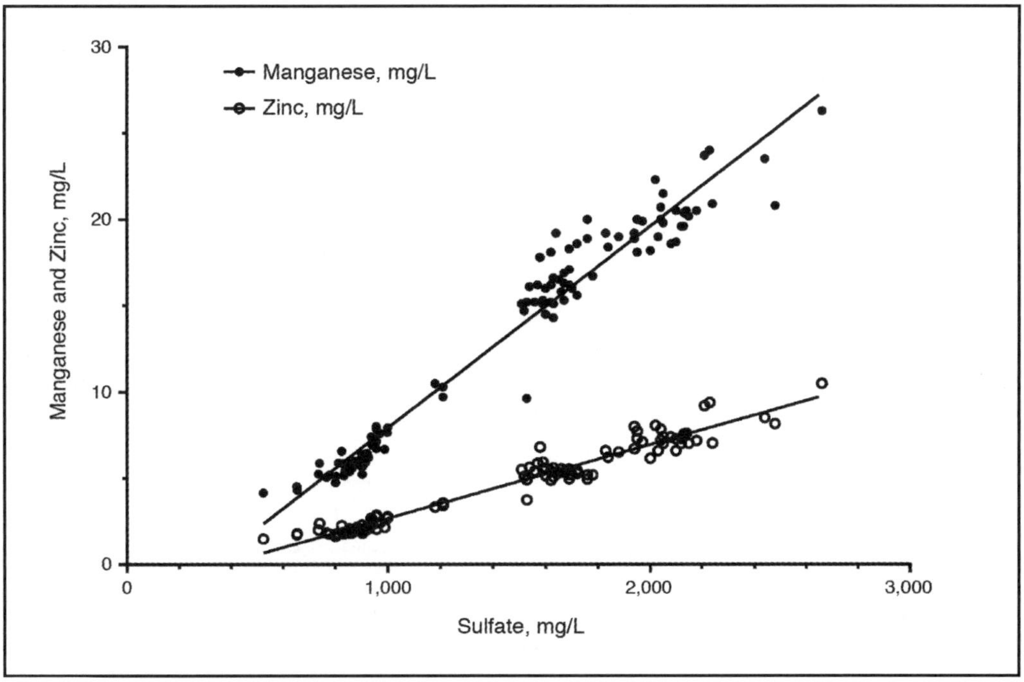

Courtesy of U.S. Geological Survey

FIGURE 9.2 Plot of Mn and Zn concentrations relative to SO_4 concentrations for acidic groundwaters in the Straight Creek natural analogue debris fan

these concentrations). Figure 9.2 shows the concentrations of Zn and Mn plotted with respect to sulfate concentrations for these acidic groundwaters. The SO_4 concentrations decreased by about 50% from top to bottom of the debris fan, which suggested more dilute waters were seeping in from the sides of the canyon during downgradient movement. The water-balance study on Straight Creek confirmed this hypothesis independently. During this downgradient dilution, these eight elements maintained a fairly constant ratio with SO_4. Aluminum behaved conservatively at low pH and nonconservatively at pH values near or above 4. This behavior can be understood in terms of pH-dependent Al hydrolysis (Nordstrom and Ball 1986). The solute correlations with SO_4 were confirmed with debris fan groundwaters that were accessible from other unmined catchments nearby and provided confidence in the analogue model.

Redox Controls

Iron concentrations did not correlate with sulfate because Fe is very sensitive to changes in both redox conditions and pH. In Straight Creek surface water, Fe was entirely oxidized, but when it appeared in the groundwater at the most upgradient well, it was predominantly reduced. Iron was reduced, but no sulfate reduction was detected. Hence, other elements whose mobility is affected by dissolved Fe reduction will also be affected. Copper concentrations did not correlate with sulfate but did correlate with the redox state of dissolved Fe, suggesting some Cu attenuation from reduction to insoluble oxides or elemental Cu. Chromium concentrations showed some decrease in anoxic waters, suggestive of reduction of Cr(6+) to insoluble Cr(3+). Redox conditions cannot be modeled in this system a priori and must be determined by analyzing the appropriate redox couples in the groundwaters. Therefore, Fe(II/III), H_2S, SO_4, and dissolved oxygen were determined. Parts of the debris fan groundwaters were mostly reduced, but other parts were not. The degree of oxidation or reduction clearly affected the solubility of redox-sensitive constituents.

Solubility Limits

Mineral solubility equilibrium could be demonstrated for calcite, siderite, rhodochrosite, amorphous aluminum hydroxide (or aluminosilicate), gypsum, and barite in circumneutral waters in the underlying volcanic aquifer, and for ferrihydrite (or schwertmannite), barite, and amorphous silica in acidic waters. These speciation and saturation index calculations were confirmed by plots of dissolved metals relative to pH, alkalinity, F, and SO_4. Hence, speciation and saturation index modeling provided a useful means of determining upper concentration limits for Fe, Mn, Al, Ca, and Ba in circumneutral groundwaters and Fe, Ba, and SiO_2 in acidic groundwaters. Mineralogical studies confirmed the assemblages found in the main lithologic units, identified mineral sources of trace elements, and correlated specific minerals with water chemistry trends.

Water and Sulfate Mass Flux in the Red River Valley

A water budget and a water catchment yield were determined for Straight Creek and also for this section of the Red River Valley (Figure 9.1). This work made it possible to estimate the flux of water and sulfate from natural ARD into the Red River. The water flux was well constrained from two synoptic surveys of the river, one in 2001 and the other in 2002. Tracer-injection studies increased the accuracy of the discharge, and all measurable inflows were determined. Sulfate concentrations were less well constrained because it was more difficult to estimate the average sulfate concentrations yielded by each of several small catchments where there might be only one well or none. Three models of groundwater–surface water mixing found that the range of sulfate flux from scar–debris fan drainages was about 900–3,000 kg/d. However, a mass balance calculation for Straight Creek drainage indicated that only 42% of the sulfate originated from pyrite weathering, with the remainder from gypsum dissolution. This sulfate mass flux was important to establish what the natural background transport of sulfate was in the Red River Valley.

Mass Balance and Reactive Transport Modeling for the Red River

The middle reach of the Red River (Figure 9.1) receives inputs of natural acidic drainage and acidic seepage from mine wastes. These acid inflows contain some Cu, Zn, Mn, and other constituents of concern. In addition to quantifying the yields of sulfate from natural acid drainage, the location where this drainage comes into the Red River and the amount of natural attenuation during downstream transport was modeled based on data collected from two synoptic sampling events with tracer injection. Synoptic sampling with discharge measurements provided the basic data for load calculations and changes in loading downriver (Kimball et al. 2002, 2006). The tracer injection allowed for more accurate and more frequent discharge measurements and, when combined with loading and other fluvial properties, made it possible to apply reactive transport models to determine downstream processes of mineral precipitation and sorption. The results of the loading modeling for the Red River showed specific locations where groundwater mixed with the river water and how much was mixing (Kimball et al. 2006). The reactive transport modeling demonstrated that, through precipitation of Fe and Al, some Cu and Zn were sorbed, thereby decreasing the dissolved concentrations of these elements.

STUDY CONCLUSIONS AND LESSONS LEARNED

A study of this type, to determine the premining groundwater chemistry at an active mine site, was quite challenging. The biggest challenge in this study was to characterize groundwater chemistry in a highly heterogeneous terrain. Mineral assemblages and their weathering rates tend to govern water composition, and these assemblages changed radically over short distances because of rapidly changing hydrothermal alteration patterns. In this study, it was important to collect extensive field data at a natural analogue site. Use of a natural analogue model was perhaps the most important example of the modeling, which is something normally not considered as modeling. Within the natural analogue model there were four other uses of models. One use was to employ saturation indices to determine if saturation was reached for mineral solubility, which established an upper limit on the concentration of potentially contaminating constituents. The second was to confirm that several elements behaved conservatively at low pH (<4), except for some redox-sensitive elements. The third was to confirm the flux of sulfuric acid from natural sources through mixing model calculations. The fourth was to apply mass loading calculations and reactive transport modeling to the distribution of contaminants to the receiving river. This downstream modeling not only allowed flux calculations to be made but also accounted for natural attenuation of trace elements during downstream transport. By maximizing the collection and interpretation of field data, there was strong scientific justification for setting limits on the range of concentrations for the constituents of concern.

In the final report (Nordstrom 2008), 10 tables of data were required to describe premining groundwater chemistry in eight catchments in which some had both acidic alluvial groundwater and circumneutral pH bedrock groundwater. Each catchment had different degrees of mineralization, which resulted in different degrees of mineralized water. Concentrations for 14 constituents of concern were usually given as a range that considered natural variability and the inherent uncertainty of the values. Many of the estimated concentrations were above the New Mexico groundwater quality standards, and these were allowed to be used in the final deliberation for closure compliance requirements for this site.

This project was initiated by the mining industry because of closure permit requirements, coordinated and reviewed through regulatory agencies and an environmental group, and executed by a third party that lacked any conflict of interest. Collaboration among all parties with regular oversight led to a successful outcome in this example, to the benefit of all parties involved.

Glynn and Brown: Application of Staged Geochemical Modeling to Understand the Past and Future of a Hydrogeochemical System

Andrew Nicholson

Application
- Modeled saturation of aquifer waters.
- Used inverse modeling to assess mineral precipitation.
- Used these results to conduct forward modeling of plume movement.
- Detailed evaluation of model sensitivity.

Codes
- PHREEQC
- PHREEQM
- NETPATH

Processes
- Dissolution of calcite by acid
- Precipitation of metals in water due to neutralization
- Mass balance concerns in modeling

Approach
- Used inverse modeling to identify precipitation phases
- Then used the inverse modeling to guide reactive transport modeling
- Applied a sensitivity analysis

Lessons Learned
- Stepwise simple to complex modeling is a good process.
- Modeling is best when based on data, including mineralogy and lab experiments.
- Model transport times were sensitive to selection of phases that precipitate.

Glynn and Brown (1996, 2012) used applied geochemical modeling to assess the transport of metal-rich, acidic, mine-derived solutions through the carbonate-containing alluvial aquifer beneath ephemeral Pinal Creek in Arizona (Figure 9.3). This mixing of the mine discharge and the groundwater resulted in the aquifer water transitioning from clean, unimpacted water; to acidified water; to a neutralized, high-sulfate water as it flowed along the aquifer. Their work was based on an ongoing U.S. Geological Survey investigation of the aquifer system and was conducted from approximately 1986 and continued to 2012 (Glynn and Brown 2012).

Pinal Creek has received acidic mine waste water from leaching processes in the Miami, Arizona, area since about 1926. These waters are characterized by elevated levels of metals (grams-per-liter levels of iron, copper, and aluminum near the sources), pH<3, high acidity, and high levels of sulfate (>50,000 mg/L). Releases of these mining solutions resulted in waters flowing through a carbonate alluvial aquifer that contains ~0.5% calcite, resulting in their neutralization. For the past 80 years, an acidic front has been moving down the Pinal Creek aquifer, having moved approximately 11 km (7 mi) downgradient of the mining area. Below the acidic front, a plume of neutralized (pH ~7) groundwater had formed with low levels of iron, copper, and other select trace metals, and high levels of manganese and sulfate.

To assess the geochemistry of Pinal Creek, Glynn and Brown took a four-step approach:

1. Evaluation of general groundwater chemistry using chemical speciation results.
2. Evaluation of mineral precipitation in the system during acidification and neutralization using inverse modeling techniques.

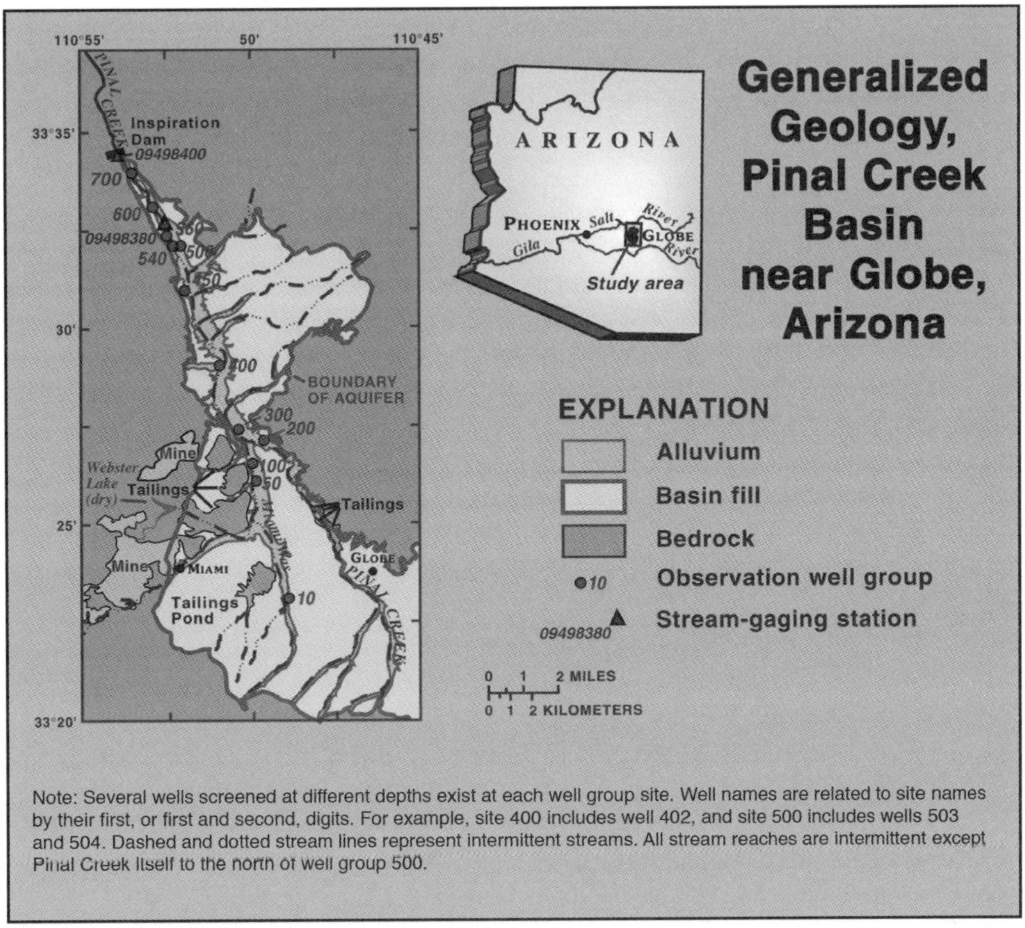

Source: Glynn and Brown 1996, courtesy of U.S. Geological Survey

FIGURE 9.3 Location and generalized geology of the Pinal Creek basin study area

3. Reactive transport modeling of the system, characterizing several geochemical processes, including calcite dissolution, gypsum precipitation, and aluminum-phase precipitation, rhodochrosite precipitation in the neutralized plume, and a sensitivity analysis by varying flow rates and dispersivity in the calculations.
4. Comparison of predicted forward modeling to field results in the Pinal Creek aquifer to assess whether the model was accurately describing the system.

The Glynn and Brown study is a relatively rare case where there is a full synthesis of field data at a mining site that includes evaluation of mineral saturation of existing waters, application of inverse modeling to assess the chemical evolution of waters during acidification and neutralization, forward modeling of the acidification and neutralization processes using coupled geochemical flow models, and a sensitivity analysis of the model. In addition, they presented the complete geochemical modeling and transport study in a single reference (Glynn and Brown 1996) that was later revised and updated (Glynn and Brown 2012).

Glynn and Brown had access to a data set that was more comprehensive than what is typically available at mining sites. Their data set included the following:

- Observation wells for piezometric measurements
- Detailed groundwater chemistry data with adequate spatial and temporal data to ensure that the existing physical and chemical systems were adequately characterized
- Detailed mineralogic analysis of aquifer materials, including detailed studies of the aquifer alteration mineralogy
- Results from column tests conducted on the aquifer material (Stollenwerk 1994), which allowed direct comparison against the model results

The geochemical model, however, was limited because it was difficult to develop a working MODFLOW-based groundwater flow model that adequately simulated the hydrologic transport in the system.

SPECIATION RESULTS

The first modeling conducted was solution speciation modeling using PHREEQC for three different waters found along the Pinal Creek aquifer flow path. This modeling served to establish the thermodynamic state of water in the system. These waters were from three wells: Well 404, an upgradient, unimpacted well; Well 504, an acidifed well in the mining-impacted plume; and Well 505, a well in the neutral plume downgradient of the site.

This speciation modeling found the following:

- The upgradient water was a calcium-carbonate, neutral-pH water, at near saturation with calcite
- The acidic plume was an iron-sulfate water, at gypsum and jarosite saturation
- The downgradient neutralized plume was a mildly acidic to neutral calcium sulfate water, near gypsum saturation, and highly supersaturated with respect to kaolinite.

This speciation modeling and characterization of the system fundamentals laid the groundwork for further evaluation of the chemical processes controlling groundwater quality.

INVERSE MODELING

Inverse modeling was conducted to simulate the acidic groundwater interacting with the carbonate-bearing aquifer, followed by its neutralization. Well 402 water interacts with the primarily clastic aquifer containing approximately 0.5% calcite. This modeling was conducted both with the codes NETPATH and PHREEQC. In this modeling, the mass of different minerals precipitating or dissolving was tracked to evaluate the processes that account for changes in groundwater chemistry. Multiple scenarios with different physical and mineralogical constraints on the system were conducted (see Table 5.6 from Glynn and Brown 1996). The choice of mineralogy and processes was based on the work of others characterizing the aquifer mineralogy (Ficklin et al. 1991) and evaluating the mobility of acidic solutions in the aquifer (Lind and Stollenwerk 1994; Stollenwerk 1994). Glynn and Brown's analysis was refined through six separate rounds of inverse modeling.

This iterative inverse modeling yielded the following conclusions:

1. Gas exsolution and dissolution were critical in simulating the system in all scenarios, specifically CO_2 exsolution.
2. Mg and Ca silicate dissolution were identified to be significant processes in the evolution of Pinal Creek aquifer waters.

TABLE 9.1 Summary of inverse modeling analyses for the Pinal Creek aquifer

Reaction or Addition*	Model 1 Tremolite Non-Minimal Model	Model 2 Tremolite	Model 3 Tremo-lite, No Calcite	Model 4 Tremolite Electron Transfer	Model 5 Tremolite $AlOHSO_4$	Model 6 Tremolite $AlOHSO_4$ Electron Transfer	Model 7 Biotite
Well 504 MF	0.258	0.258	0.258	0.258	0.258	0.258	0.258
Well 504 MF minimum	0.258	0.258	0.258	0.258	0.258	0.258	0.258
Well 504 MF maximum	0.277	0.277	0.277	0.277	0.264	0.264	0.263
Dolomite +	0.398		2.191		0.290		2.291
Gypsum –							
Goethite –	–8.292	–8.292	–8.292	–8.291	–8.292	–8.292	–9.074
Calcite +	3.588	4.383		1.626	3.817	3.991	
Anorthite +			2.512	2.202			2.691
$AlOHSO_4$ –					–0.423	–0.423	
Gibbsite –	–0.423	–0.423	–5.448	–4.827			–2.167
SiO_2 –	–3.649	–3.649	–5.803	–8.687	–3.165	–3.629	
Rhodochrosite –	–2.757	–2.757	–2.756		–2.769	–2.364	–3.253
MnO_2 +	3.903	3.903	3.903	6.659	3.903	4.309	4.294
$Mn(OH)_3$ –				–5.512		–0.811	
Tremolite +	0.500	0.500	0.142	0.580	0.440	0.498	
Biotite +							0.521
K-montmorillonite –							–1.796
Net protons consumed	–0.279	–0.281	–0.272	–0.266	–0.271	–0.270	–0.298
Alkalinity change	0.214	0.213	0.213	0.213	0.213	0.213	0.190
difference	0.493	0.494	0.485	0.479	0.484	0.483	0.487

Courtesy of U.S. Geological Survey
*MF, mixing fraction; +, dissolution only; –, precipitation only

3. Oxidation of iron by manganese oxides was a likely mechanism for the lowering of iron concentrations and the increase in manganese concentrations.
4. To account for the manganese mass balance, it was necessary to invoke the precipitation of a reduced, Mn^{2+}, manganese phase.
5. Several different combinations of mineral phases could explain the aluminum/silica/iron chemical evolution of the system.

This inverse modeling directed the geochemical and mineralogical constraints to be implemented in further, prospective reactive transport modeling, as shown in Table 9.1.

REACTIVE TRANSPORT MODELING

Reactive transport modeling was applied to evaluate the chemical evolution of the system. The modeling allowed Glynn and Brown to assess how different mineralogical and geochemical constraints affected the rate at which the acidic front advanced. Intrinsic to their analysis were some additional assumptions and understanding of the system based on findings from other studies in the Pinal Creek system, including those related to

- The groundwater velocity field,
- The amount of dilution and other transport processes, and
- Additional detail on the composition of the aquifer material.

The reactive transport modeling was conducted in a one-dimensional fashion using PHREEQM and later PHREEQC, which had incorporated the necessary transport code routines. The mobility of the low-pH front in the aquifer was assessed by varying several specific variables to assess the sensitivity of the system. These included

- The solid carbonate concentration;
- Adding longitudinal dispersivity;
- Including or excluding dolomite;
- Including or excluding rhodochrosite precipitation;
- Varying the possible aluminum phases to precipitate, including $Al(OH)_3$, kaolinite, or $AlOHSO_4$;
- Varying partial CO_2 pressures;
- Including or excluding cation exchange;
- Including or excluding surface-complexation-based sorption;
- Including or excluding irreversible dissolution of Ca and Mg silicate minerals.

Reaction path modeling runs were conducted assuming a 5.6-km- (3.5-mi-) long column, divided into 10 cells. The average groundwater rate was assumed to be 5 m/d (16.4 ft/d).

The speed of the acidic front moving downstream was highly dependent on the amount of carbonate in the system (Figure 9.4). The modeling results showed that pH plateaus were commonly created due to the formation of mineralogical buffers. The pH was not expected to decrease monotonically through time.

In addition to the amount of calcite in the system, the speed of the acidic front was also sensitive to the ratio of carbonate to the amount of MnO_2 (CM ratio) in the system, with the plume moving faster when the CM ratio was lower. Also, the choice of Al-bearing phases in the system affected the results. Incorporation of $AlOHSO_4$ instead of $Al(OH)_3$ resulted in slower plume movement, consistent with the findings of other workers (e.g., Nicholson et al. 2003).

More extensive forward reactive transport modeling was done and the effects of redox and Al chemistry on solute transport tested in the most recent paper (Glynn and Brown 2012).

COMPARISON TO FIELD CONDITIONS

The estimated retardation of the acidic plume (the rate of movement of the groundwater divided by the rate of movement of the acidic front) based on field observation ranged from 5 to 8.4. Model results ranged from 3.2 to 8.1, with a typical value of 4.7 for these simulations. The pH plateaus determined in the results of modeling studies were within the pH ranges observed in the field.

The results obtained by Glynn and Brown show that even with detailed system characterization, there can be uncertainty in specific geochemical mechanisms that control the transport of aqueous species in an aquifer. However, they effectively simulated the behavior of the Pinal Creek plume within the constraints of the system. Their model can be used to constrain the expected behavior in the Pinal Creek aquifer and aide in management decisions for the site.

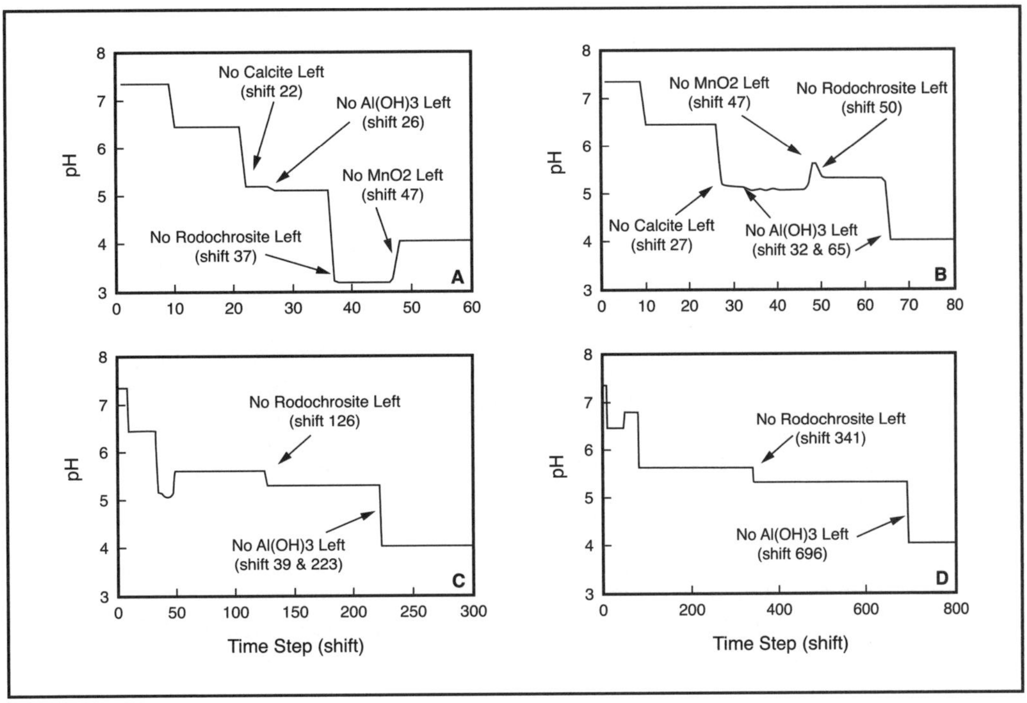

Courtesy of U.S. Geological Survey

FIGURE 9.4 Curves for pH breakthrough of four PHREEQC advection-reaction simulations that use the basic reaction model (see Glynn and Brown 1996, 2012) with amorphous $Al(OH)_3$ as the Al-bearing phase allowed to precipitate. No dolomite was included. Initial calcite concentrations for each of the simulations shown: (a) 2.1×10^{-2} mol/kg H_2O, (b) 3.0×10^{-2} mol/kg H_2O, (c) 4.2×10^{-2} mol/kg H_2O, and (d) 8.4×10^{-2} mol/kg H_2O.

LESSONS LEARNED

The work of Glynn and Brown provides several specific insights into the process and applicability of geochemical modeling:

- **Simple to complex.** By approaching the problem stepwise, from simpler mass balance and inverse modeling to more complex reactive transport modeling, they were able to identify system controls that would likely not have been identified had they gone directly to more complex modeling.
- **The importance of data.** The authors based their modeling on both temporal field data (detailed mineralogical studies of the aquifer materials) and laboratory simulations of aquifer acidification and neutralization. This approach aided in constraining the modeling assumptions.
- **Mineralogy and thermodynamic data matter.** Glynn and Brown found that the choice of minerals precipitating and the assumed solubility of those minerals strongly affected the rate of transport in the aquifer. Careful consideration of this factor is essential in environmental systems because metastable minerals, with variable solubility due to poorly formed crystal structures, can dominate mine-water impacted systems.

Hydrologic Modeling

Walter Weinig

Application

- Modeled water quality impacts for a proposed phosphate mine.
- Evaluated different capping alternatives of overburden material.
- Calibrated to regional groundwater elevation and surface water flows data.

Codes

- MODFLOW-SURFACT

Processes

- Unsaturated zone infiltration through cap
- Groundwater–surface water interaction
- Groundwater flow in a complex geologic setting

Approach

- Developed overarching conceptual site model.
- Used unsaturated zone leaching model.
- Calibrated to steady state and transient state.

Lessons Learned

- Modeling facilitates alternative analyses.
- Code selection is guided by conceptual site model.
- Public domain code is allowed for transparency during permitting.

BLACKFOOT BRIDGE IMPACT ANALYSIS

Song and Leone (2013) summarized a hydrologic modeling application to evaluate potential impacts to water quality from proposed phosphate-mining activities in Idaho. The proposed mining project could affect surface-water and groundwater resources through release of metals and other constituents into the environment. Results of the modeling study were used to support the environmental impact statement (EIS) for the proposed project and provided input to the mine design for permitting.

BACKGROUND

The proposed mine consists of an open pit to be developed in three panels, plus auxiliary mine facilities. Mining activities are expected to occur over 17 years once development begins.

The stratigraphic section in the project area includes Quaternary-age alluvium, colluvium, and travertine deposits; Tertiary-age basalt flows, tuff, and rhyolite; and Mississippian- to Triassic-age sedimentary deposits. The structural geology is complex. A major fault discharges deeply circulating groundwater to a series of springs and wetlands along the mountain front, affecting regional groundwater flow. A series of high-angle normal faults cut through two of the planned open pit panels. With one exception, displacement of bedding across the fault planes is minor.

Several streams and surface-water bodies occur within or near the project area. The major surface-water feature is the Blackfoot River, which flows along the northern property boundary (Figure 9.5). The Blackfoot River is designated as a Special Resource Water under State of Idaho water quality standards, recognized as needing protection to maintain outstanding characteristics and current beneficial uses.

MODEL CODE SELECTION AND MODEL CONSTRUCTION

A hydrologic model was developed using the variably saturated groundwater flow and solute transport code MODFLOW-SURFACT. MODFLOW-SURFACT uses a finite-difference discretization scheme to numerically solve complex flow and transport problems. The code was

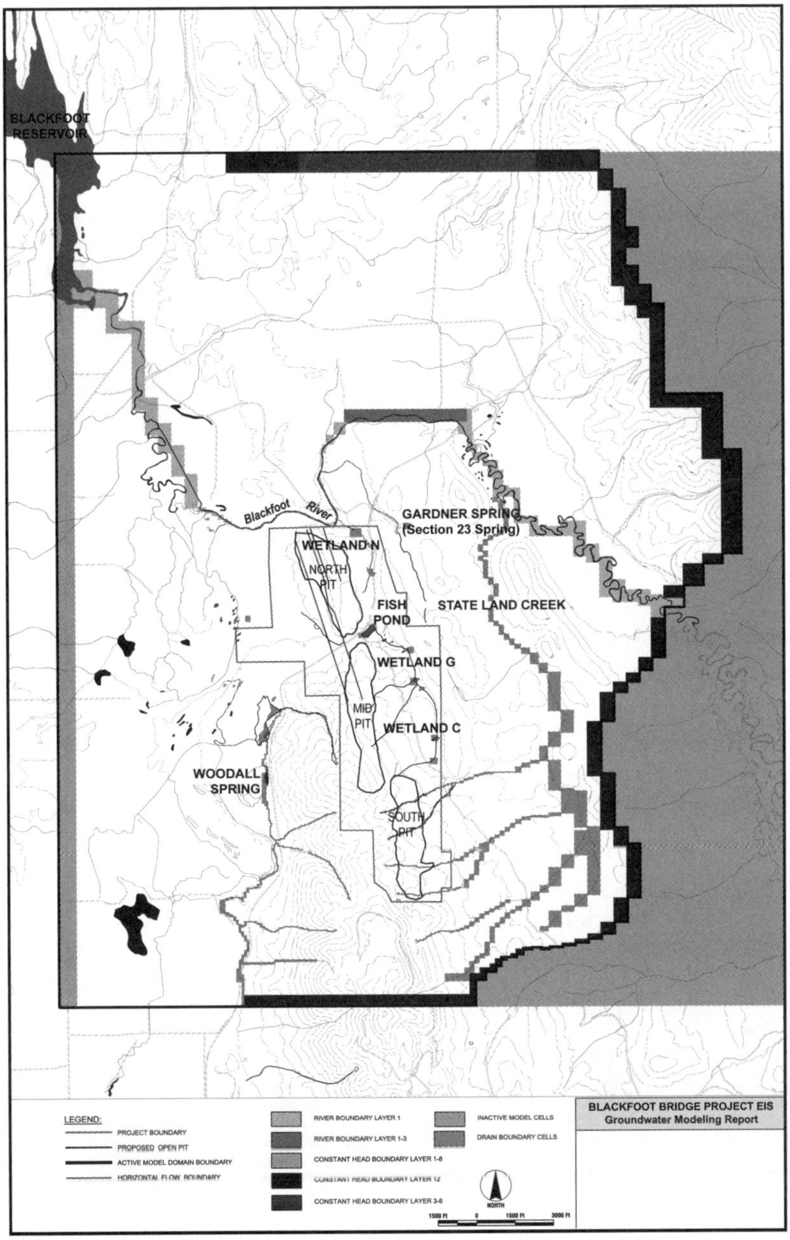

Source: Song and Leone 2013

FIGURE 9.5 Model domain and boundary conditions

selected because of its ability to simulate steep hydraulic gradients and cell rewetting, and for its advanced numerical capabilities with respect to solving the transport equations.

The model domain (Figure 9.5) covered an area of about 70 km^2 (27 mi^2). Boundaries of the model grid were set at a significant distance from the project area to minimize the influence of boundary conditions on simulation results in the area of interest.

The finite-difference grid was divided into 204 columns, 400 rows, and 12 layers. Song and Leone (2013) indicate that horizontal grid spacing was variable based on the size of the area of interest, the total area of the model domain, and the degree of accuracy and precision needed. Model layers were designed to represent the significant hydrostratigraphic units identified in the conceptual site model and the development of the mine pits.

FLOW MODEL CALIBRATION

According to Song and Leone (2013), the hydrologic model was initially calibrated to steady-state conditions. Two different transient calibrations were then performed: one based on seasonal variations in recharge and a second based on the results of an aquifer test. After the transient calibrations were complete, the steady-state model was rerun to confirm that a single set of parameter values satisfied both the steady-state and transient targets.

GEOCHEMICAL SOURCE TERMS

Results of site-specific geochemical characterization studies (Whetstone Associates 2010) were used to evaluate the geochemical characteristics and leaching behaviors of the overburden and ore that would be produced from the proposed mine. Selenium was selected as the constituent of concern for the site. Laboratory columns were prepared and leaching tests were conducted over several leaching cycles to obtain leachate concentrations for all potential source areas.

Results of the geochemical characterization indicated that the most significant potential source of selenium contributing to loading in groundwater would be leachate from unsaturated source areas (Song and Leone 2013). The actual amount of selenium potentially discharging to groundwater would also depend on the net percolation rate through different cover types applied over the source areas.

HYDROLOGIC MODEL RESULTS

Song and Leone (2013) indicate that three development scenarios were evaluated using the hydrologic model: the Proposed Action, Alternative 1A, and Alternative 1B. Three types of cover systems were considered in the three scenarios: the Base Case, the Simple 1, and the Complex 2.

According to Song and Leone (2013), the Proposed Action scenario included a combination of the Base Case cover system and the Simple 1 cover system. Alternative 1A involved placing a combination of the Complex 2 cover system and the Simple 1 cover system over the various source areas. Alternative 1B replaced a portion of the Simple 1 cover from Alternative 1A with the Complex 2 cover system, resulting in the use of the Complex 2 cover system over nearly 90% of the source areas.

Simulation results for Alternative 1A as presented by Song and Leone (2013) are shown in Figures 9.6 and 9.7. Figure 9.6 presents the simulated maximum selenium plume in groundwater due to mining activities while Figure 9.7 shows the anticipated impact of selenium to surface water in the vicinity of the project. A sensitivity analysis was performed to evaluate uncertainties related to two key parameters: the hydraulic properties of one of the aquifers and the unsaturated and saturated source strengths.

LESSONS LEARNED

Song and Leone (2013) summarized the results of a modeling study performed to support preparation of an EIS and mine design and permitting activities. The Song and Leone study illustrates some important aspects of the use of hydrologic models:

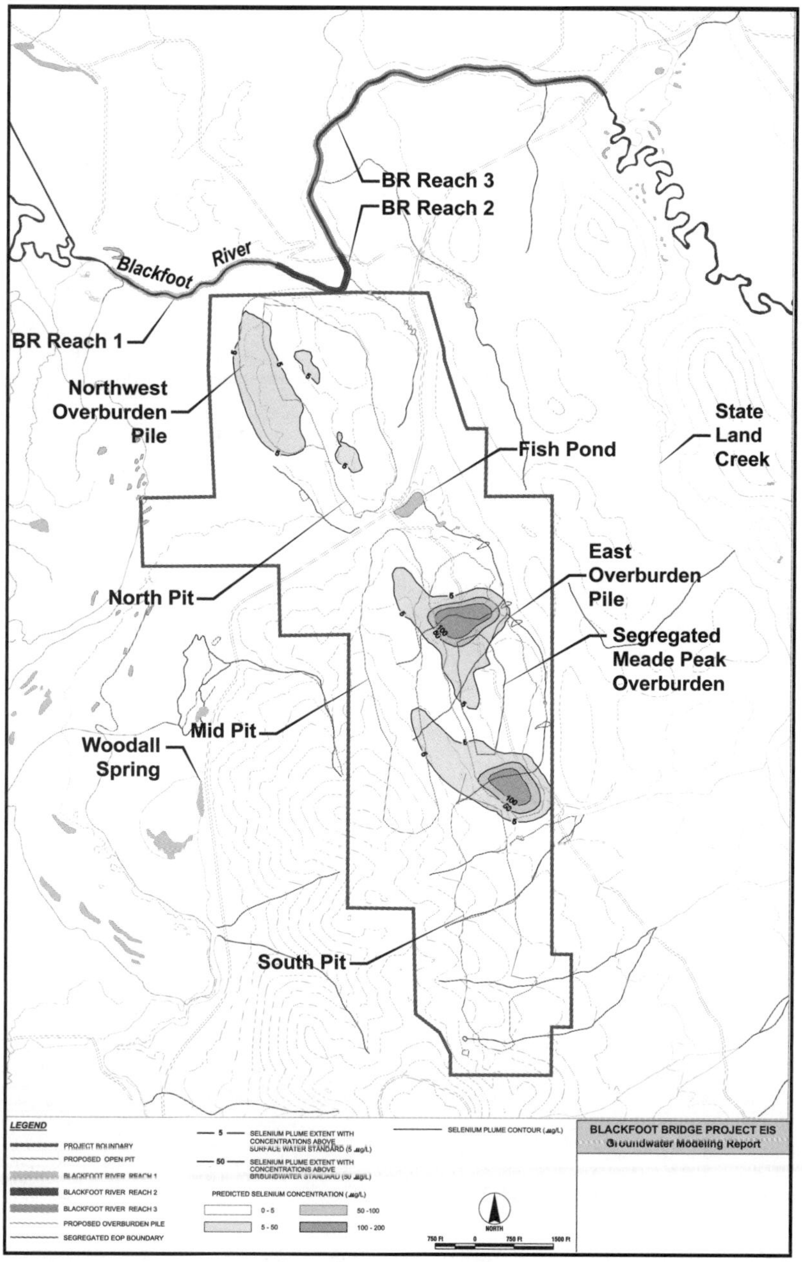

Source: Song and Leone 2013

FIGURE 9.6 Simulated selenium extent in Alternative 1A

- **Hydrologic modeling helps facilitate evaluation of alternatives.** As Song and Leone (2013) point out, numerical modeling and quantitative analysis are necessary to facilitate mining activity evaluation and assess alternatives to mitigate potentially unacceptable impacts to water resources. The hydrologic model allowed assessment of several alternative scenarios using a common conceptual and numerical framework.
- **Model codes should be selected based on project-specific requirements.** The computer code used in this study (MODFLOW-SURFACT) was selected after the conceptual site

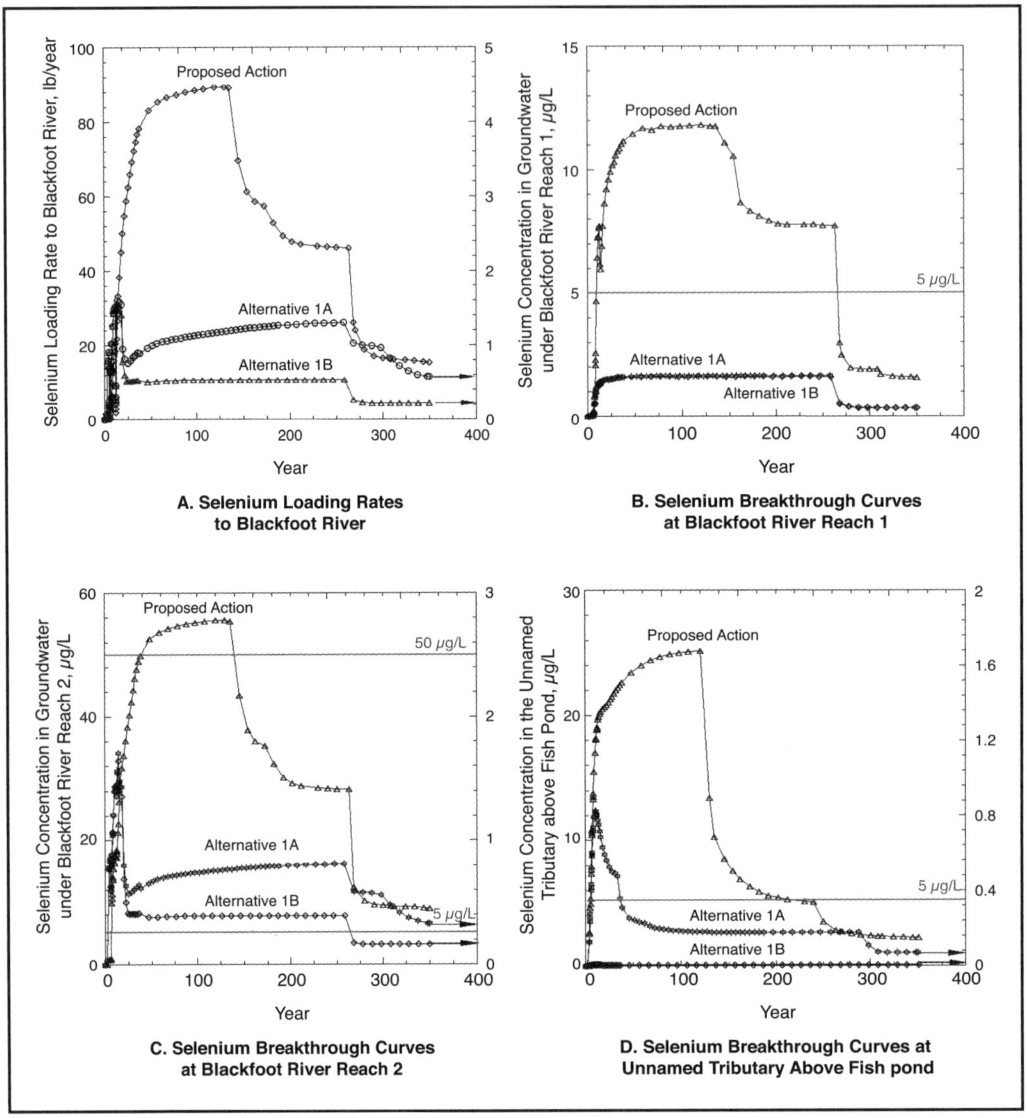

Source: Song and Leone 2013

FIGURE 9.7 Simulated selenium impact to surface water

model was developed and the critical requirements for output were defined. As illustrated in this handbook, a large number of codes and modeling approaches are available for any particular problem. Selection of an appropriate code should be based on project needs and available data.

- **Transparency is important.** In the highly regulated environment typical of a proposed mining project, the modeling approach and underlying data should be transparent and available for review by all stakeholders. Use of a well-documented and widely available code helps facilitate review of model results and assessment of their validity for the problem at hand. Proprietary and undocumented codes should not be selected when detailed review by regulators or external stakeholders is expected.

Buckhorn Mine Groundwater Flow and Contaminant Transport Modeling

Ann Maest

Application

- Prediction of groundwater elevations during mine operations
- Prediction of nitrate concentration trends in nearby stream

Codes

- FEFLOW
- FracMan

Processes

- Effect of faults on groundwater flow directions and elevations
- Effect of groundwater pumping on water table

Approach

- Develop conceptual model of groundwater flow system.
- Incorporate identified faults, measured groundwater elevations, and streamflow.
- Calibrate to measured groundwater elevations, streamflow, and water quality tracer study results.
- Use calibrated model to estimate size and extent of capture zone, effects of dewatering wells, and groundwater flow paths.

Lessons Learned

- Faults control movement of groundwater and mine constituents.
- Groundwater flow and contaminant transport modeling are highly uncertain in a fractured bedrock setting.

BACKGROUND

The Buckhorn mine is an underground gold mine in north-central Washington State, less than 16 km (10 mi) from the Canadian border (Figure 9.8). Mining began in January 2008. Gold ore is extracted using block caving methods, and all ore is trucked off-site for beneficiation and processing at the Kettle River mill and tailings facility, approximately 80 km (50 mi) to the southeast. Waste rock is temporarily stored on-site and placed as cemented or uncemented backfill in the underground mine as mining progresses. The ore host and surrounding rocks are comprised of altered and unaltered andesite and clastics (including volcaniclastics), skarns, marble, and unmineralized rhyolitic/granodioritic dikes (Washington Department of Ecology 2005). The two lobes of the mine are the shallower Southwest zone, where mining first began, and the deeper Gold Bowl zone to the northeast, which has a higher sulfide content. The ore is hosted primarily in garnet and magnetite skarns, and the most common sulfide mineral is pyrrhotite. The mine is located at the head of three drainages: Nicholson, Myers, and Marias Creeks. Most groundwater moves along faults. Dewatering wells and sumps are used to dewater the mine.

MODEL DEVELOPMENT AND CALIBRATION

A transient FEFLOW model for the mine was developed by URS Corporation (2006) and was used in the Final Supplemental Environmental Impact Statement (Washington Department of Ecology 2006) to predict groundwater inflows to the mine and changes to groundwater elevations and streamflows as a result of mine operations. The model was updated by Golder Associates Inc. (2011) to include additional faults, dewatering wells, streamflow, groundwater elevation data, monitoring wells, and discharge locations (Golder Associates Inc. 2011). The FracMan code was used to help input the faults to the FEFLOW mesh and allow for easy manipulation of data and checking for consistency. The hydraulic conductivity (K) values used for the faults in the updated

FIGURE 9.8 Overview map of Buckhorn mine location

model were based on older and more recent pump tests. Precipitation and recharge measurements were derived largely from the Republic meteorological station, which typically had approximately 10 cm (4 in.) less precipitation than the Buckhorn mine station (available only from 2009–2010 for the model). The timing of precipitation is also different at the Republic and Buckhorn stations because of the differences in elevation and location.

The updated model was calibrated to groundwater elevation data from six monitoring wells and base flows from 13 surface water monitoring stations using data from 2006 to 2010. The calibrated model was evaluated against data from 1990 to 2005. Modifications in structure and fault properties were made to the calibrated 2006 model to reach an acceptable match with measured adit flows and groundwater elevations in the updated model. The best matches with water elevations and streamflows were attained using a hydraulic conductivity value of 15.2 cm/d (0.5 ft/d) for all faults except NWF-2, which used a K value of 61 cm/d (2.0 ft/d).

MODEL RESULTS AND COMPARISON TO OPERATIONAL CONDITIONS

The calibrated model results for two monitoring locations are compared to measured stream discharge and groundwater elevations in Figure 9.9. Although the model is a transient model, it does not adequately capture the peaks in stream discharge associated with snowmelt each spring. Groundwater elevations show more variability seasonally, but modeled peaks are not well matched with measured peak water levels. The relatively low K values assigned to the faults could help explain the discrepancies. Although the model used values of 15.2 and 61 cm/d (0.5 and 2.0 ft/d) for the faults, another report estimated hydraulic conductivity of the North Lookout Fault (NLF) zone at 12.2–26 m/d (40–85 ft/d). Under snowmelt and groundwater recharge conditions experienced in May and early June 2011, the model was numerically unstable (Golder Associates Inc. 2011). During this snowmelt and snowmelt conditions in other years, water levels in the fractured bedrock rose by up to 61 m (200 ft), and the model could not converge. The

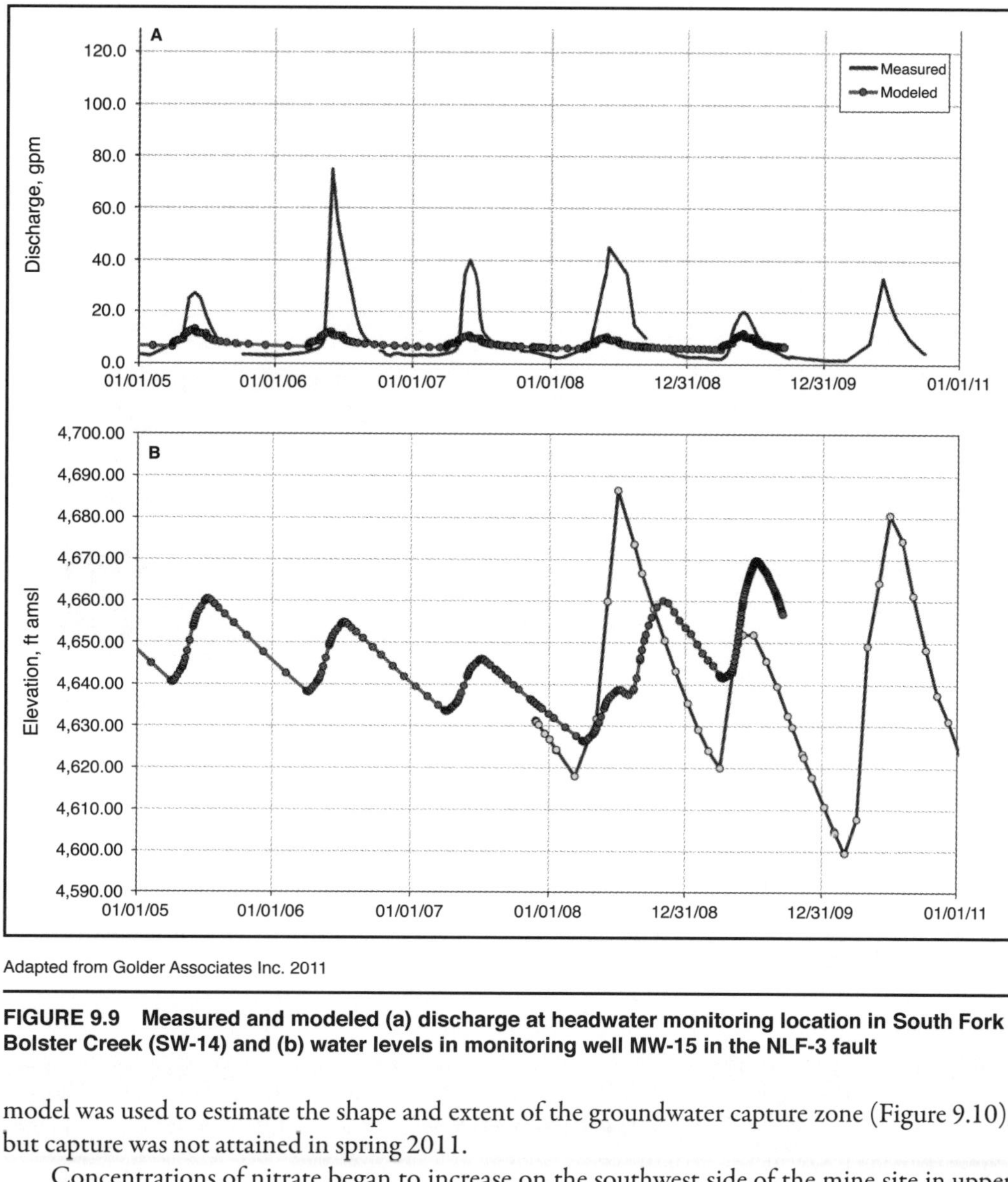

Adapted from Golder Associates Inc. 2011

FIGURE 9.9 Measured and modeled (a) discharge at headwater monitoring location in South Fork Bolster Creek (SW-14) and (b) water levels in monitoring well MW-15 in the NLF-3 fault

model was used to estimate the shape and extent of the groundwater capture zone (Figure 9.10), but capture was not attained in spring 2011.

Concentrations of nitrate began to increase on the southwest side of the mine site in upper South Fork Bolster Creek in 2008. Nitrate concentrations can be elevated at mine sites from blasting agents used to extract the ore. Possible sources of high-nitrate water include mine water in the underground mine and infiltrated runoff from waste rock. The FEFLOW model added nitrate as a tracer at possible source locations to help understand the transport to South Fork Bolster Creek (SW-14). Measured and modeled nitrate concentrations at SW-14 are shown in Figures 9.11 and 9.12. The modeled concentrations were based on an assumption that the source is water stored in a large sump in the underground mine. The model predicted that nitrate concentrations would peak in early spring runoff from 2008 to 2011 but would decrease each year after 2009; in 2012, concentrations were predicted to return to approximate baseline conditions. Actual nitrate concentrations instead increased from 2008 through 2011 and have not returned to baseline values (~0.5 mg/L as N at this location).

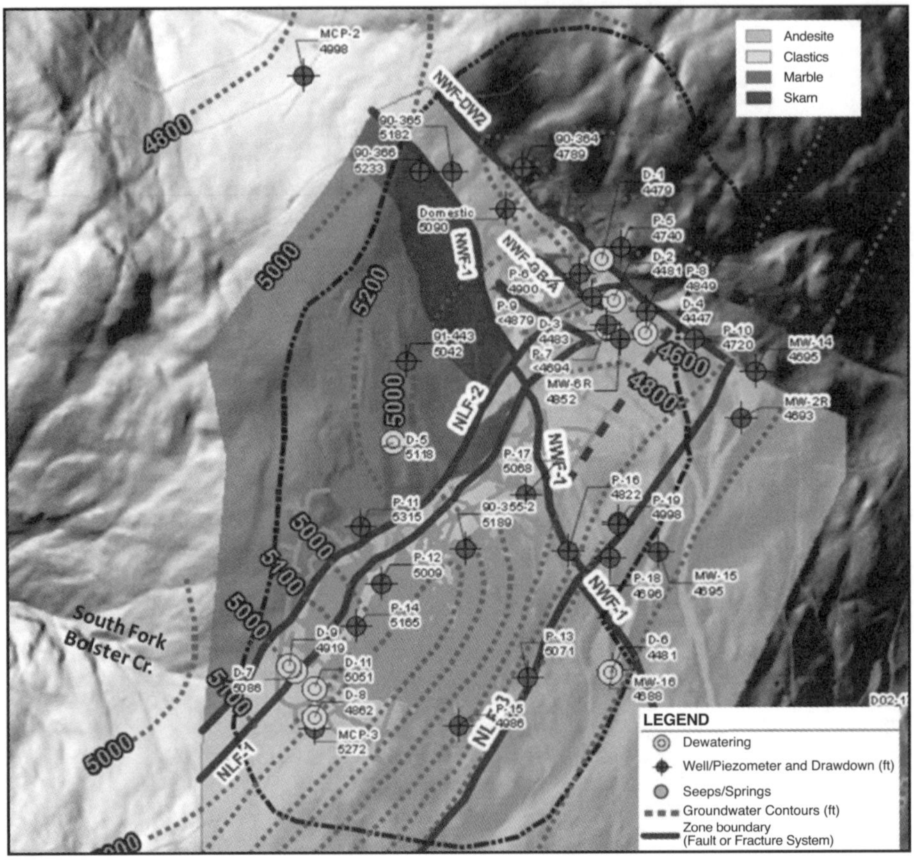

Adapted from Golder Associates Inc. 2014

FIGURE 9.10 Map of modeled capture zone (black dash-dot line), monitoring wells and piezometers, faults (thick lines), and groundwater elevations (dotted lines)

SUMMARY

The FEFLOW model for the Buckhorn mine is not able to simulate site conditions, including leakage of mine-related constituents to streams, discharge or groundwater elevations during snowmelt, or streamflows throughout the year. Fractured bedrock is difficult to model, and more work is needed to simulate actual site conditions.

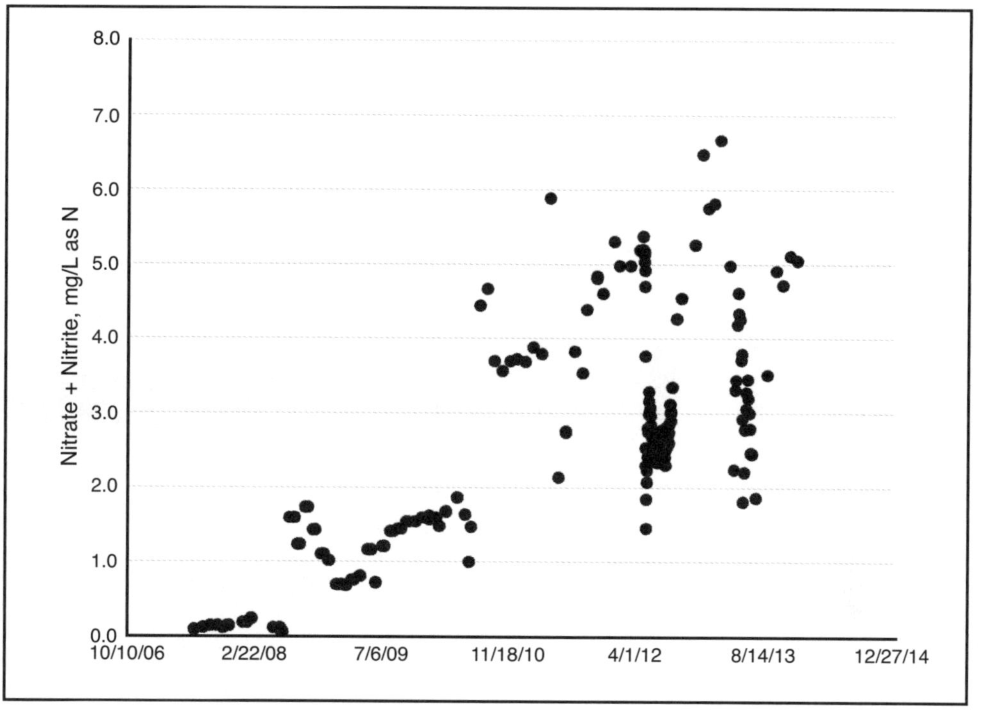

Data from Washington Department of Ecology 2007–2013

FIGURE 9.11 Measured nitrate concentrations in South Fork Bolster Creek (SW-14), 2007–2013

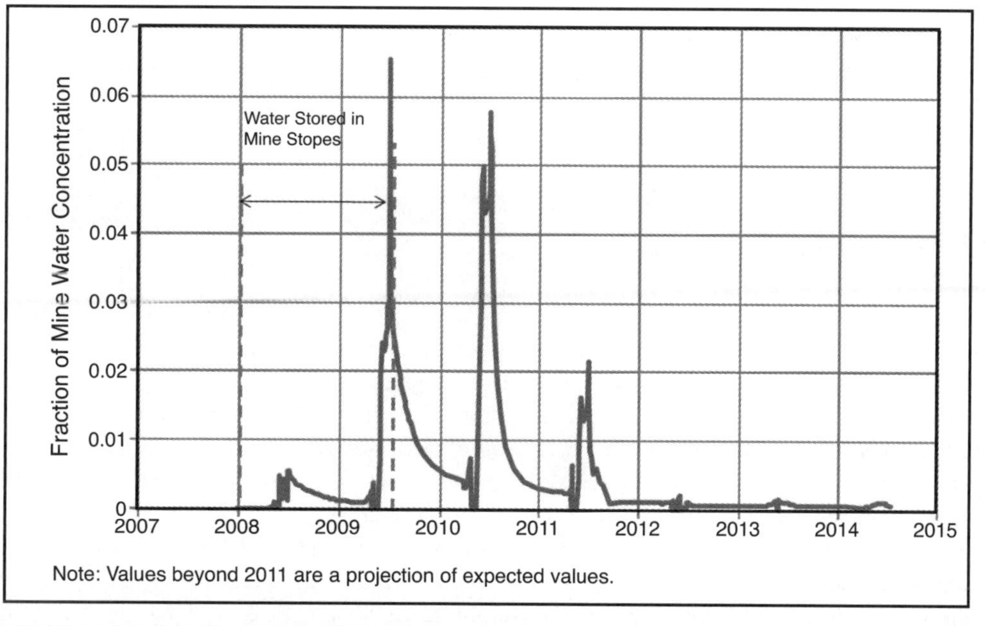

Source: Adapted from Golder Associates Inc. 2011

FIGURE 9.12 Modeled nitrate concentrations, as a fraction of mine water concentration, in South Fork Bolster Creek (SW-14)

Uranium Processing Wastes and Groundwater Plumes: Three Field Studies

D. Kirk Nordstrom

Uranium wastes in the form of waste rock, tailings, and further processed wastes will oxidize and weather to form soluble uranium and other soluble contaminants. In a mineral deposit, uranium commonly occurs in a reduced form, U(IV), which is very insoluble. Upon oxidation to U(VI), it becomes quite soluble, especially when complexed with carbonate ions. Two recent reviews describe the occurrence and mineralogy of uranium, the mining and mineral processing, and the biogeochemical processes affecting mobility (Newsome et al. 2014; Campbell et al. 2015).

The three case studies described here involve reactive transport modeling from uranium wastes. The first is about the sulfuric acid leachate mobility at Riverton, Wyoming; the second is about uranium reactive transport in the shallow aquifer of the Naturita waste site, Colorado; and the third is about reactive transport for bioremediation at the Rifle waste site, Colorado.

Uranium Study I: Riverton, Wyoming

Application

- Modeled groundwater flow from a uranium-contaminated site
- Incorporated dissolution and precipitation of major minerals (calcite, gypsum, gibbsite, ferric hydroxide)

Codes

- TRUMP
- PHREEQE
- DYNAMIX

Processes

- Calcite dissolution
- Gypsum precipitation
- Gibbsite precipitation
- Ferric hydroxide precipitation
- Groundwtater mixing
- Acid neutralization
- CO_2 production

Approach

- Finite difference method for groundwater flow
- Considered advective and diffusive flow

Lessons Learned

- Groundwater flow models can be effectively combined with geochemical models to characterize rate and mechanism of contaminant transport from a uranium waste site.
- Proportion of tailings water contaminating groundwater can be estimated.

STUDY I: RIVERTON, WYOMING

During the processing of uranium, sulfuric acid solutions are typically used to leach uranium from ore that has been finely ground in a mill. After the uranium has been extracted, some of that sulfuric acid becomes waste. The waste slurry containing fine sands was deposited in evaporation ponds without liners. At Riverton, Fremont County, Wyoming, about 1 billion kg of processed ore and spent leach solutions were deposited on Quaternary deposits that overlie coarse sandstones. Narasimhan et al. (1986) modeled the groundwater flow for this site with the TRUMP code and coupled it with the PHREEQE code to form the DYNAMIX code. Infiltration by both initial processing fluids and rainwater, saturated and unsaturated groundwater flow, and chemical reactions all had to be considered. The acid was neutralized by calcite occurring in the aquifer, promoting the precipitation of gypsum and aluminum and iron hydroxides (White et al. 1984). An estimated 1.7% of the process tailings water had mixed with the natural groundwater. It would appear that attenuation processes, such as precipitation and sorption, would be more important in controlling contaminant concentrations than dispersion, but the authors explicitly stated that they did not do any sensitivity or uncertainty analyses to determine that likelihood. They also did not attempt to predict future groundwater contamination scenarios because additional work would be required to confirm the relative importance of the various attenuation processes. In spite of this study's shortcomings, it was a pioneering effort at the time and it demonstrated what reactive transport modeling was possible, how it could help with site characterization, and what the potential was for predicting plume migration.

Uranium Study II: Naturita, Colorado

Application

- Simulated uranium transport and sorption in groundwater with a semi-mechanistic method
- Wide range of uranium and alkalinity concentrations

Codes

- FITEQL
- MODFLOW
- MT3DMS

Processes

- U(VI) adsorption
- Calcite saturation

Approach

- Used a surface complexation model (SCM) to simulate K_D values

Lessons Learned

- Sorption modeling can account for attenuation of U(VI) during groundwater flow with highly variable uranium and alkalinity concentrations.
- There was improvement over just using a constant KD without SCM.
- Sensitivity of sorption is highly sensitive to recharge rate and to alkalinitiest.

STUDY II: NATURITA, COLORADO

The former uranium and vanadium mill site near Naturita, Colorado, processed these elements from the 1930s until the mill was shut down in 1958, except for uranium concentrates that were produced but not processed at this site from 1961–1963. After salt roasting, both carbonate and sulfuric acid leaches were used for processing. Mill tailings and contaminated soils were removed. Groundwaters in a shallow alluvial aquifer became contaminated from leached uranium. The groundwater was modeled using MODFLOW, and nonreactive transport was simulated with the code MT3DMS. Hydraulic conductivity was estimated using slug tests, isotopic age dating, and chloride transport. Investigations by Curtis et al. (2006) simulated U(VI) transport of the contaminated groundwaters using a surface complexation model (SCM) and the FITEQL code. Simulated K_D (distribution coefficient) values using SCM varied strongly with highly variable uranium and alkalinity concentrations because of a variable recharge rate. Saturation indices indicated equilibrium solubility with respect to calcite and undersaturation with respect to oxidized uranium minerals, uraninite, and U_3O_8. Batch adsorption experiments with sediments from the site provided data of a general nature that were fit without reference to specific mineral surfaces and used postulated reaction stoichiometries based on typical secondary coatings of soils and sediments. The sorption model could account for uranium attenuation without having detailed knowledge of the minerals and their surfaces.

STUDY III: RIFLE, COLORADO

At several uranium-contaminated sites, usually mill tailings sites, reductive bioremediation has been examined as a possible strategy to immobilize groundwater uranium plumes. By promoting metal-reducing bacteria with injections of soluble organics, such as acetate, soluble U(VI) would be reduced to insoluble U(IV). The creation of anoxic conditions would keep the uranium stabilized in a solid phase. Reduction might occur either directly by microbial enzymatic reactions or indirectly by reduction from other components such as Fe(II) produced by microbial iron reduction.

Near Rifle, Colorado, uranium mill tailings (now removed) have leached into the groundwater table, raising uranium concentrations to 0.1–0.4 mg/L. The site is on a flood bank of the Colorado River with an unconfined alluvial aquifer. Several bioremediation experiments have been studied from 2002 to 2008. Installation of several wells with long-term monitoring provided the hydrologic, chemical, mineralogical, and microbial data necessary to model the field site. The site was investigated not only with a series of repeated injections of organics, each succeeding experiment improving over the previous, but also with the development of more complex models, including the evolution from one-dimensional (1-D) to three-dimensional (3-D) models (Fang et al. 2009; Yabusaki et al. 2011). The primary result of several years of investigations was that microbial stimulation required larger quantities of organic carbon loading than anticipated. When iron reduction increases, dissolved uranium decreases, but when sulfate reduction begins, increases in U(VI) are seen because the organic is consumed in this process and iron and uranium reduction decreases. Organic concentrations must be about triple than what is necessary to sustain sulfate reduction if iron and uranium reduction is to continue. Other results include the recognition of preferential flow paths that greatly affect the temporal and spatial rates of reductive processes as well as flow velocities and mineral dissolution and precipitation. Modeling at this site not only made it possible to interpret the results of complex hydrobiogeochemical processes, but it also provided much-needed guidance during the evolution of the field experiments over six-plus years.

Uranium Study III: Rifle, Colorado

Application

- 3-D variably saturated flow coupled to biogeochemical reactive transport

Codes

- eSTOMP on a massively parallel computer

Processes

- U(VI) adsorption/desorption
- U(VI), Fe(III), and S(VI) reduction
- Microbial growth
- Mineral dissolution and precipitation (uraninite, siderite, calcite, S°, FeS, Fe_3O_4, and FeOOH)

Approach

- Used equilibrium constants from the Nuclear Energy Agency database (Grenthe et al. 1992).
- Used rate constants from the literature.

Lessons Learned

- Uranium reduction is promoted by enhancement of anoxic microbial activity through organic carbon injections.
- The amount of organic carbon needed to promote U(VI) reduction must substantially exceed that necessary to maintain sulfate reduction.
- Modeling 3-D variable flow in a physically and chemically heterogeneous system is computationally demanding.

Simulations of Reactive Transport Modeling in Surface Waters

D. Kirk Nordstrom

Runkel et al. (1996) developed a reactive transport model for trace metal contaminants in mountainous stream systems. The first code for this purpose was OTIS (One-Dimensional Transport with Inflow and Storage; Runkel 1998). The code was coupled with the hydrochemical code, MINTEQA2 (Allison et al. 1991), and the resulting code became OTEQ (Runkel 2010), which can calculate reactive transport models. The OTEQ code has been used extensively for a large number of mountainous streams contaminated with acid mine drainage (AMD) for purposes of characterizing and remediating these contaminated surface waters (e.g., see those cited in Nordstrom 2011).

The four case studies depicted here describe simulations done with these codes in a testing capacity. The first case reports on a field study in which sulfuric acid was added to a stream containing AMD (the Snake River near Montezuma, Colorado), and downstream monitoring was done to compare with simulations. The second explains simulations done for a pH neutralization field experiment. The third describes how a calibrated reactive transport model can be used to test the consequences of different remediation scenarios on downstream water chemistry. The fourth is a more severe test, called a postaudit, and involves simulating transport of several trace elements during pH neutralization without knowing the field results until after the simulation is completed. Initial calibration is done with field data and then the simulation is conducted blind except for knowledge of base injection. Reactive transport modeling for surface waters has the advantage of more direct access to the system under study than similar modeling for subsurface systems. Furthermore, some of the conclusions derived from surface-water reactive transport modeling should have implications for improved subsurface system modeling.

Simulation Study I: Decrease in pH, Snake River, Colorado

Application

- Modeled changes in pH and Fe concentrations as a result of adding sulfuric acid to a mildly acidic stream.
- Incorporated dissolution and precipitation of ferric hydroxide with mixing.

Codes

- OTEQ (in an earlier form)
- OTIS (in an earlier form)
- MINTEQA2

Processes

- Acidification
- Ferric hydroxide dissolution
- Iron photoreduction
- Mixing

Approach

- Constant tracer injection was used to define hydraulic properties.
- Changes in pH and iron concentrations were measured.
- Simulations were obtained with reactive transport model based on calibration to background conditions before adding acid.

Lessons Learned

- Equilibrium-based reactive transport satisfactorily simulated the field results.
- The ion activity product for $Fe(OH)_3$ varied little.
- Iron solubility could be modeled, assuming equilibria for moderately acid conditions.

STUDY I: SNAKE RIVER, COLORADO

For 3 hours, a constant injection of sulfuric acid was added to the Snake River, a naturally acidic stream in Colorado (McKnight and Bencala 1989). A lithium chloride solution was also injected for 6 hours as a conservative tracer to determine discharge and other hydraulic transport parameters. The pH decreased from 4.2 to 3.2 at the first of five sampling stations downstream from the injection point over a total distance of 2.8 km (1.7 mi). Flows are typically 0.1–0.5 m^3/s. The decrease in pH caused an increase in iron concentrations to a constant level that could be modeled with a solubility product constant of $Fe(OH)_3$ of pK_{sp} = 40.2–40.8. Variations in Fe concentration were caused by dissolution of freshly precipitated ferric hydroxide and photoreduction. Only Fe concentrations and pH varied significantly during the acidification experiment. An equilibrium-based reactive transport model was applied to the data and was found to satisfactorily simulate the iron dissolution process and pH changes (Runkel et al. 1996). Most other solutes were found to behave conservatively as would be expected for an acid water. This was the first successful application of a surface-water model for AMD incorporated into a code (that became OTEQ) and an important step forward in reactive transport modeling of contaminated streams.

Simulation Study II: Increase in pH, St. Kevin Gulch, Colorado

Application

- Modeled changes in pH, Fe, and Al concentrations after adding $NaHCO_3$ to an acidic stream.
- Incorporated oxidation, hydrolysis, and precipitation of ferric hydroxide with mixing.

Codes

- OTEQ (in an earlier form)
- OTIS (in an earlier form)
- MINTEQA2

Processes

- Neutralization and mixing
- Ferric hydroxide precipitation
- Fe photoreduction
- Al precipitation
- Mixing

Approach

- Constant tracer injection with sodium was used to define hydraulic parameters.
- Changes in pH, Fe, and Al concentrations were measured.
- Simulations were obtained with reactive transport model.

Lessons Learned

- Equilibrium-based reactive transport was unsatisfactory.
- Kinetic constraints delayed oxidation and precipitation of Fe.
- Dissolved Al was affected by streambed interactions.

STUDY II: ST. KEVIN GULCH, COLORADO

A 500-m (1,640-ft) reach of AMD in St. Kevin Gulch, Colorado, was used for a field experiment on neutralization (Broshears et al. 1996). The results were exploited to test the simulation capability of a reactive transport model. Prior to the pH modification experiment, a LiCl tracer constant-injection study obtained discharges and other hydraulic parameters. Constant injection of a concentrated $NaHCO_3$ solution occurred over a 5.6-hour period in two steps, and water sampling was collected from just upstream of the injection point and four stations downstream. The pH increased from 3.5 to 4.2 in the first step and to 5.8 in the second step at the first downstream station (24 m [79 ft] downstream from the injection). However, at the most downstream station (498 m [1,634 ft] from injection), the pH only increased to 4.5.

The interpretation of the results was that the oxidation of ferrous iron was slow compared to the water transport rate and could not be modeled by instantaneous oxidation and equilibrium

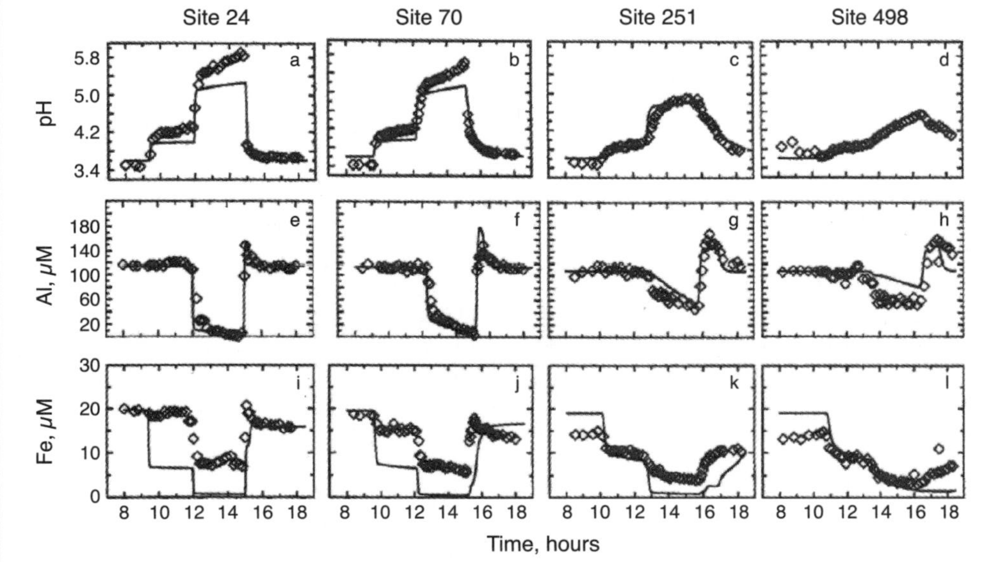

Note: Data points represent measured values, and the solid lines are the simulations.

Source: Broshears et al. 1996, courtesy of U.S. Geological Survey

FIGURE 9.13 Measurements of pH, and dissolved Al and Fe concentrations during a pH neutralization experiment

precipitation. Slow ferrous iron oxidation results in higher pH and Fe concentrations than simulated for the sites nearer the injection point. Further downstream, the effects of slow iron oxidation are less pronounced, and modeled and measured pH and Fe gradually converge (Figure 9.13). Determinations of dissolved ferrous and ferric concentrations were not made in this study, so they were estimated based on improved fits between simulations and measurements for both Fe concentrations and pH. These reactive processes were further complicated by rates of photoreduction, sorption reactions, and precipitate dissolution (when neutralization stopped). Aluminum gave better results in the simulation than Fe and pH. The measurements showed that the main mismatch occurred when neutralization stopped and aluminum concentrations rebounded when there was an initial spike in Al. Freshly precipitated Al redissolved when the pH decreased to give this spike (Figure 9.13). For this study, equilibrium-based modeling gave poor comparisons between simulation and measurement, and the processes whose rates were needed to improve the simulations were identified and discussed.

STUDY III: USING A REACTIVE TRANSPORT MODEL TO EVALUATE REMEDIAL ALTERNATIVES

If a catchment or drainage basin containing several inputs of acid mine drainage (portal effluent, tailings and waste rock seeps, open pit drainage) can be synoptically sampled for contaminants with a constant injection of a tracer for more reliable discharge estimates, then it is possible to calibrate a reactive transport model and then simulate conditions under different remediation scenarios to estimate changes in the resultant water chemistry. In other words, the consequences of different combinations of remedial alternatives can be simulated to see what has the greatest effect on improving water quality before making the final management decisions. Sometimes a combination of remedial actions and natural attenuation can have the most cost-effective solution for remediation. Such an application has been done with successful results for the Summitville Superfund mine site in Colorado (Ball et al. 2004). The first report on this approach was from Runkel and Kimball (2002) for Mineral Creek, San Juan Mountains, Colorado. Mineral Creek is a 15-km (9.3-mi) tributary of the Animas River (where an accidental mine water release contaminated surface-water supplies in four states in early August of 2015). The OTEQ code was used for simulating baseline steady-state conditions, and inputs included streamflow from tracer dilution, synoptic sampling for downstream chemistry, inflow chemistry and discharge of tributaries and seeps, and mass action expressions for dissolution/precipitation reactions, sorption reactions, and complexation reactions.

The dominant features of the downstream chemical profiles were reproduced along a 3.5-km (2.2-mi) reach in which the main processes were precipitation of Fe(III) and Al hydroxides and sorption of Cu, As, and Pb onto freshly precipitating Fe(III) hydroxides (using the surface complexation model and database of Dzombak and Morel 1990). Additional simulations were then run under two different remediation scenarios. Both simulations dosed a small acidic stream of pH 2.4 with $CaCO_3$ to a pH of ~7. In the first simulation, the treatment system was closed so that Fe(II) was not oxidized, and in the second simulation, the treatment system was open such that the Fe was oxidized to Fe(III). Both scenarios increased stream pH and substantially reduced total and dissolved concentrations of Al, As, Cu, and Fe by the end of the study reach. Dissolved concentrations of Pb were reduced much more under the second Fe-oxidized scenario and Zn, Ni, and Cd were affected very little.

Simulation Study III: Evaluating Remedial Options for Mineral Creek, San Juan Mountains, Colorado

Application

- Modeled changes in pH and metal concentrations after two hypothetical remedial options.
- Incorporated oxidation, hydrolysis, and precipitation of ferric hydroxide with mixing.

Code

- OTEQ

Processes

- Neutralization and mixing
- Ferric hydroxide precipitation
- Al precipitation
- Trace metal sorption (Cu, Pb, Zn, Cd, As, Ni)
- Mixing

Approach

- Constant tracer injection with synoptic sampling
- Simulations obtained with reactive transport model by adding $CaCO_3$ under two treatment conditions: (1) Fe(II) kept reduced and (2) Fe(II) oxidized during treatment

Lessons Learned

- Calibration to current conditions was successful.
- The cost and time necessary to complete the study for complex terrains can be substantial.

Although this study shows that this reactive transport model is a promising tool, several limitations were discussed. The importance of highly reliable streamflow, inflow chemistry, and in-stream water chemistry was stressed. Collecting this data requires considerable investment in personnel, equipment, and water analyses. If the data are compromised, the uncertainty can escalate substantially and may not be meaningful for treatment design. Mass balances can be further compromised if base flows add sizable contributions to the loading and these cannot be measured. Seasonal and climatic variations in flows will change the mass balances, and repeating the synoptic sampling with tracer injection is required to determine the range of conditions for highly variable flows. Finally, error propagation from analytics, streamflow parameters, and thermodynamic data need to be demonstrated to add confidence in the results, and sensitivity analysis would be useful for identifying the most critical parameters as accurately as possible. An examination of errors for the trace metal loading of Peru Creek, Colorado, using the same tracer injection with replicate synoptic sampling was provided by Runkel et al. (2013). The lower error of tracer dilution to estimate discharges compared to standard acoustic Doppler measurements was demonstrated, although it must be noted that tracer dilution does not work for a losing stream. Otherwise, replication gave considerable confidence to the reactive transport modeling because other sources of error were within reasonable bounds. This modeling has proven useful for systems of complex hydrology and geochemistry in mountainous terrains.

Simulation Study IV: A Model Postaudit, Mineral Creek, San Juan Mountains, Colorado

Application

- Predicted changes in pH and metal concentrations during base addition to an acid drainage with minimal input data and compared to measured values.
- Based solely on pre-neutralization conditions and initial base addition.

Code

- OTEQ
- MINTEQ

Processes

- Neutralization and mixing
- Ferric hydroxide precipitation
- Al precipitation
- Trace metal sorption (Cu, Pb, Zn, Cd, As, Mn)
- Mixing

Approach

- Constant tracer injection with synoptic sampling calibration for pre-remediation
- Given initial condition for base addition

Lessons Learned

- Prediction was generally favorable and useful for regulatory purposes.
- A major source of error was estimating the mass of fresh precipitate to sorb trace metals.
- Another source of error was the surface complexation constant for sorption of As.

STUDY IV: A MODEL POSTAUDIT—BLIND SIMULATION OF NEUTRALIZATION

A postaudit study for reactive transport modeling on a 2-km (1.2-mi) acidic reach of Mineral Creek, Colorado (pH = 3.1 to 4.8), provided an opportunity to predict how well a model could simulate the effects of acid neutralization (Runkel et al. 2012). After initial calibration using OTEQ under low-flow steady-state conditions, a simulation was run with the only known variables from the base neutralization study, which were changes in streamflow and the pH and alkalinity at injection (after mixing). Simulations with MINTEQ were run to develop revised concentrations for dissolved inorganic carbon, total hydrogen ion, and Fe(II) and Fe(III) concentrations at the upstream boundary condition. Precipitation reactions were limited to ferrihydrite and gibbsite. Gas exchange was not considered; equilibrium with atmospheric CO_2 was assumed throughout. Two different sets of surface complexation constants were considered. Also, when precipitates formed, they could remain in the water column or they could partially settle out. This possible range was considered.

The results from the postaudit were presented in terms of percent relative error (measured concentration, minus simulated concentration, times 100, divided by measured concentration). A second metric was used to determine how well the simulation predicted attainment of water quality standards. In general, the predictions agreed well for the most downstream compliance point. Predictions for Al, As, Fe, H^+, and Pb were better than those for Zn, Cd, Cu, and Mn.

Because of the difficulty in estimating the mass (and surface area) of precipitated Fe, there tended to be an underestimation of sorbed Zn, Cd, Cu, and Pb. Sorption of As was overestimated and likely is related to the surface complexation constant.

APPENDIX A

Recap: What Models Can and Cannot Do

D. Kirk Nordstrom

This appendix is a modified and expanded version of "Recommended Guidelines on the Use of Models" from Nordstrom (2012).

1. **A conceptual model is critical to the success or failure of any geochemical modeling.** How we conceive of a site and what data we collect to improve our conception of a site is more important than any other modeling we do.
2. **Modeling cannot give an exact answer, only an approximation.** Modeling is often, but not always, performed with mathematics. Mathematics gives exact answers to idealistic and simplistic questions. Pure mathematics can give a unique answer to a question, but it is not science and it is not realistic. Modeling, when carefully utilized, can constrain the possibilities and even approximate the probabilities, but uniqueness is not a modeling characteristic unless the problem is simple, and then modeling may not be needed.
3. **Modeling can never substitute for reliable and relevant field data.** All modeling, when applied to a specific site, depends on the quality of site characterization. The more sophisticated and extensive the modeling and the greater the paucity of the field data, the greater the likelihood the model is not adequately constrained. With little field data, it becomes easier to model geochemical processes so that any desired result can be produced. *Corollary:* The more high-quality field data that are acquired, the more constraints available for modeling and the fewer assumptions needed. Ask yourself whether the field observations and data drive the modeling or vice versa. Model output, as always, is directly dependent on the data and assumptions that go into the model. Collecting too much field data is also possible, and the model should be checked for the most sensitive parameters and only data that will most constrain the likely models should be collected.
4. **The largest weaknesses of any model computation are the quality of the input data and the adequacy of the assumptions (implicit and explicit).** There are generally two explicit parts to the input data: the field data and the thermodynamic and/or kinetic databases. Both are common sources of error that get propagated in model computations. Field data are most commonly collected by low-salaried workers who often do not know or appreciate the implications of mistakes or errors in field measurements and are not motivated to record problems or errors with field work. There are dozens of possible databases that could be used for thermodynamic data. Many of them contain internal inconsistencies that may or may not propagate serious errors in the output. If preferred databases are not being used, it would be prudent to test the database by comparing the results when the database is revised to one of the preferred databases.

   Assumptions used in a model computation should always be clearly stated. Some assumptions are often not articulated because they are considered to be well established or widely known. An example would be the assumption of rapid equilibrium for aqueous solution-phase speciation. Chemical measurements on solution reaction equilibria have established this assumption as true for 99% of reactions occurring in natural waters. There are a few that do not, but these are rare and unusual situations, mostly having to do with the formation of complex or polymeric species whose equilibrium constants are questionably reversible. Solubility equilibria, however, typically are not attained quickly and depend on the rate of reaction relative to the flow rate of the water. Hence, solubility equilibrium is more likely to be reached in aquifers with slow velocities rather than in surface waters.

5. **Model computations are not unique.** The fewer the constraints, the wider the range of possible computational results. Anyone who claims that his or her results are unique is not being honest. The best that one can hope for is to have enough constraints to narrow the possible results to the one (or more) outcome that is the most likely.
6. **Model and code reliability can be tested in some limited ways.** Independent evaluation of the reliability of the model computation gives the user or interpreter more confidence in the computational output. There is some analogy here to validation of water analyses by quality assurance/quality control protocols. One example is given previously where the effect of changing the database can be considered to determine what the effect is on the output. Another example is to perform a computation on a water analysis that has undergone a round-robin speciation computation by several codes, as in Nordstrom et al. (1979). This exercise would take only a few minutes for any modeler, but it would be revealing if the results were not reasonably consistent. Other examples and test cases can be found in Ball and Nordstrom (1991), Parkhurst and Appelo (1999), Parkhurst (1997), Zhu and Anderson (2002), and Nordstrom and Campbell (2014). Ultimately, we are looking for an independent measure of the system performance. The only other reliable measure that works is "expert opinion." That is why careful and thorough independent review is essential.
7. **The main conclusion or argument based on model computation should be reproducible in a simpler manner by hand calculation.** Results from model computations should be reproducible, in a simpler way, such as through a back-of-the-envelope calculation. If a modeler cannot evaluate a computation or argument based on computations by a hand calculation or a spreadsheet computation and get the same basic results, then there must be something wrong (with either the computation or the hand calculation).
8. **Model computations must be explicable to non-modelers.** If the results cannot be clearly explained to non-modelers who are familiar with the basic physical and chemical principles involved, then it raises the question whether the modeler really understands them.
9. **No matter how much data are acquired, no matter how sophisticated the modeling, there are always factors one cannot know that limit our ability to predict the future.** Our ability to predict and control the world around us is inherently limited. Uncertainty stems from chaotic sources that are both natural and human derived. The environment is much richer in complexity and uncertainty than any of our models can accommodate.
10. **The more sophisticated the modeling, the less we know about how well the model performs or how it works (the "complexity paradox").** Oreskes (2000a) has an excellent description of this problem: "The closer a model comes to a full representation of a complex earth system, the harder it is to evaluate it. There is a trade-off between representation and refutability."

11. **Is it necessary to predict?** An important question that is rarely asked: What did we do before modeling? We made measurements and did calculations to estimate the nature of the system under investigation—it was similar, but we did not call it modeling. Oreskes (2000b) points out that, historically, earth scientists did not usually make predictions. Instead, they studied the earth by studying current processes and looking at the rock record. Further, model evaluation should focus as much on the input data as on the output and how the output compares to independent observational data. Indeed, some predictions are meaningless and unnecessary to achieve a particular regulatory goal.
12. **Transparency is essential.** Proprietary codes have no place in a regulatory arena. An independent programmer and modeler must have access to the code to determine if it really does what is claimed by its sponsor.
13. **Develop a model evaluation plan.** Have a set of criteria and a plan for how to evaluate computational results from a given code. Propagate errors and uncertainties where possible. A sensitivity analysis can help determine what factors, variables, or parameters have the greatest influence on the results. An uncertainty analysis can determine what are the most important sources of uncertainty and whether they can be estimated or not.
14. **Choose a code that has been tested and is appropriate for the particular objectives required.** There are numerous geochemical codes, and one should consider using a code that is well suited and adequately tested for the particular site conditions under investigation.
15. **Remedial alternatives can be evaluated more quantitatively through the use of codes that can simulate the consequences (simulation scenarios) of different potential options.** Once a contaminant site has been adequately characterized, simulation scenarios can be compared for a variety of remedial options to optimize the remediation in terms of the most cost-effective approach.

APPENDIX B

Fifteen-Year Retrospection on Glynn and Brown (1996)

D. Kirk Nordstrom

The series of papers by Pierre Glynn and/or Jim Brown on the Pinal Creek, Globe–Miami, site in Arizona provides the most extensive and comprehensive application of geochemical models to a mine waste site that the authors of this handbook are aware of. The last paper Glynn and Brown published regarding this site (Glynn and Brown 2012) includes the following 15-year retrospection by senior author Glynn with some general thoughts on how best to model contaminated mine sites that require hydrogeological as well as geochemical interpretations.

THE SENIOR AUTHOR'S 15-YEAR PERSPECTIVE ON THE GLYNN AND BROWN (1996) PAPER

The geochemical modeling conducted by Glynn and Brown (1996), while confirming and refining knowledge about chemical reactions in the Pinal Creek groundwater system, also highlighted significant knowledge gaps. The Glynn and Brown (1996) study provided a basis for further groundwater geochemical investigations at the Pinal Creek site, including several reactive transport modeling studies. For example, Brown et al. (1998) modified and refined the simulation of groundwater evolution presented in Brown (1996) for 1-D flow (with lateral dilutions as needed) from Well 51 to Well 701 from 1984 to 1994. Some of the reactions were changed in the modified model as well as the spatial distribution of those reactions. For example, rhodochrosite was allowed to precipitate and CO_2 was allowed to equilibrate with a pCO_2 of $10^{-1.33}$ (typical of the unsaturated zone) in the "neutralized" zone downgradient from Well 503. Sensitivity analyses examined, for example, surface complexation and sorption of metals on iron oxyhydroxide. Sorption was relatively less important than initially considered by the authors in controlling pH and dissolved metal concentrations relative to other processes (such as redox processes, carbonate dissolution, and dilution). Longitudinal hydrodynamic dispersion was found to be relatively unimportant (a beta version of PHREEQC v.2 was used). Generally, good fits were obtained for the simulation of heavy metals such as Cu, Zn, and Ni. However, the simulation of pH and dissolved Fe and Mn still proved problematic, and a simulation of the carbonate and redox reaction kinetics was still needed.

Brown et al. (2000), again improving on Brown (1996), found that HYDROGEOCHEM and PHREEQC simulations of the Stollenwerk (1994) column experiments provided similar numerical results, but once again, required accounting for reaction kinetics, if the experimental results were to be matched by the numerical simulations. In the absence of adequate knowledge of reaction rates, the observed experimental results could also have been matched by fitting irreversible mass transfers of water and reactants in PHREEQC. A common result of the Brown et al. (2000, 1998) studies was that the numerical simulation results were reasonable only if the number of adsorption sites was reduced to 5% of the initial value considered by Stollenwerk (1994). Stollenwerk simulated, through the use of the geochemical code MINTEQA2, the neutralization and later rinse-out of Pinal Creek acidic groundwater that he observed in a column experiment. Because MINTEQA2 was not a reactive transport code, Stollenwerk used a series of "batch" calculations to simulate transport, and explicitly defined the aqueous and solid-phase chemical inputs for each cell at each time step. As mentioned in Brown et al. (2000), "Although output from one time step provided much of the input for the next time step, the aqueous and solid-phase concentrations of selected constituents were manually adjusted to fit measured concentrations in column outflow and measured changes in column mineralogy."

To improve the characterisation of calcite, dolomite, and pyrolusite (MnO_2) dissolution rates at the Pinal Creek site, Brown and Glynn (2003) conducted in situ field experiments that suspended well-characterized commercially obtained mineral samples in acidic waters at the Pinal Creek site (Wells 101 and 301) for various lengths of time (from 96 to 595 days) starting in April 1998. In situ dissolution rates for calcite and dolomite were determined, and found to be about three orders of magnitude slower than rates determined by Plummer et al. (1978) and Busenberg and Plummer (1982) in laboratory experiments. Interestingly, Brown and Glynn (2003) found that the incorporation of the in situ rates, or alternatively, the faster laboratory-determined rates, in PHREEQC v.2 (Parkhurst and Appelo 1999) reactive transport simulations did not result in better fits of Stollenwerk's column experiments. Instead, an equilibrium model adequately simulated carbonate dissolution at the shorter and faster (relative to the field situation) spatial and temporal scales of the column experiment. More surprisingly, Brown and Glynn (2003) observed growth in the mass of the pyrolusite material placed in the in situ experiments, despite thermodynamic conditions that, in theory, favored pyrolusite dissolution. They postulated that this result could be related to the result of Villinski et al. (2001), who found that the dissolution of MnO_2 under similar conditions in a laboratory flow-through cell resulted in the precipitation of a mixed oxidation Mn-Fe phase with the structure of jacobsite ($MnFe_2O_4$). Brown and Glynn (2003) then used the rate equations developed by Postma and Appelo (2000) for the reductive dissolution of birnessite (MnO_2) by dissolved Fe(II) to better fit the distribution and evolution of dissolved Mn and Fe concentrations in the transition zone and in the acidic part of the plume (at well 402) from 1984 through 1992. Although the rate constant used in their PHREEQC v.2 simulations was about 3 to 4 orders of magnitude smaller than the range of values used by Postma and Appelo (2000) in their column experiments, Brown and Glynn (2003) did obtain a better fit of the observed Fe, and especially, the observed Mn concentrations in the transition zone. Their simulated pH values, however, remained about 1 pH unit higher than the observed values.

The field, laboratory and numerical modeling investigations of the Pinal Creek basin provide several lessons that may be useful to other investigations of complex, highly dynamic, systems with a relative paucity of information (and limited funding to obtain information):

1. Constructing, analyzing and interpreting numerical models, regardless of the type of model (hydrologic vs. geochemical; inverse vs. forward), forces the modeler(s), and hopefully the user (s) of the models, to reexamine and revise their

conceptual model and perceptions of the available information. The modeling process forces the modelers and users to assemble, structure, transform, and assess a wide variety of information.

2. No model is ever final. As numerical modeling leads to greater understanding of a system, it invariably illuminates knowledge gaps and the need for more information. Frequent iteration is needed among field observation, field experiments, laboratory experiments, and the periodic assembly and interpretation of the available information through numerical modeling.
3. The combination of inverse modeling (including inverse geochemical modeling) and forward or "predictive" modeling is particularly powerful in helping assess available information. Inverse modeling forces the modeler to use basic principles (such as conservation of mass and energy) to *interpret* the available observations in terms of properties of the system (e.g., geochemical mass transfers, possible reactions, other intrinsic system properties). Forward, or predictive modeling, forces the modeler to *test* available information by "extrapolating" forward through time, space or other informational dimensions by using the modeler's preconceptions of system behavior. In the case of the Pinal Creek system, inverse modeling helped Glynn and Brown (1996) identify sets of reactions that could explain the observed geochemical evolution of groundwaters, and helped identify field and laboratory observations/experiments that might improve characterization and quantification of active geochemical processes. Sensitivity analyses conducted through "forward" reactive transport modeling identified which minerals and reaction processes, given knowledge of reaction thermodynamics and stoichiometries, provided reasonable rates of migration of chemical fronts and a reasonable distribution of solution properties (such as pH and Fe and Mn concentrations) in space and time. More importantly, the geochemical modeling at the Pinal Creek site helped identify which reactions and processes were *unreasonable*.
4. Determining what is *unreasonable* is generally dependent on limited information and observations (through time, space and process domains) and on an *interpretation* of that limited information. In complex, highly dynamic, systems with limited regularity or steadiness of observations, such as the Pinal Creek groundwater system, the problem of eliminating unreasonable processes is exacerbated: what might seem reasonable, or unreasonable, given available information at some point in time and space could change. The studies conducted at the Pinal Creek site illustrate the fact that nature always keeps surprises in reserve for its observers and interpreters. Humility, and frequent testing of assumptions, are needed in modeling nature's systems.
5. Given our often limited knowledge of natural systems, it behooves us to model these systems by considering *general* system behavior before interpreting, matching, and predicting *specific* system behavior. Curve and point-by-point matching of numerical simulations to the observed chemical evolution at the Pinal Creek site, or laboratory columns, was generally difficult and provided limited, albeit useful, returns on expended effort. Inverse modeling and sensitivity analyses of general system behavior, however, through relatively unconstrained reactive transport simulations, provided substantial insights on knowledge gaps and

information needs that could improve understanding of the complex, highly dynamic processes at the site. These modeling efforts should have been conducted much earlier in the life of the Pinal Creek project investigations.

6. Finally, the Pinal Creek project taught us the value of spatial and temporal observations. Systems that have a complex suite of active processes and that exhibit dynamic changes in observed properties require extensive (and often costly) adaptive monitoring programs that have sufficient resolution in both time and space. Considering a greater variety of geochemical/hydrological information than might initially be considered in setting up a monitoring program is useful and can sometimes help compensate for a lack of spatial/temporal monitoring: different geochemical or hydrological measurements often reflect different degrees of spatial integration and temporal evolution.

References

Acero, P., Ayora, C., Carrera, J., Saaltink, M.W., and Olivella, S. 2009. Multiphase flow and reactive transport model in vadose tailings. *Appl. Geochem.* 24(7):1238–1250.

Albarède, F. 1995. *Introduction to Geochemical Modeling*. Cambridge: Cambridge University Press.

Allison, J.D., Brown, D.S., and Novo-Gradac, K.J. 1991. *MINTEQA2/PRODEFA2, A Geochemical Assessment Model for Environmental Systems: Version 3.0 User's Manual*. EPA/600/3-91/021. Athens, GA: U.S. Environmental Protection Agency.

Alpers, C.A., and Nordstrom, D.K. 1999. Geochemical modeling of water–rock interactions in mining environments. In *The Environmental Geochemistry of Mineral Deposits. Reviews in Economic Geology*, Vol. 6A. Edited by G.S. Plumlee and M.J. Logsdon. Littleton, CO: Society of Economic Geologists. pp. 289–234.

Alpers, C.N., Nordstrom, D.K., and Spitzley, J. 2003. Extreme acid mine drainage from a pyritic massive sulfide deposit: The Iron Mountain end-member. In *Environmental Aspects of Mine Wastes*, Vol. 31. Edited by J.L. Jambor, D.W. Blowes, and A.I.M. Ritchie. Ottawa, ON: Mineralogical Association of Canada. pp. 407–430.

Amos, R.T., Blowes, D.L., Smith, L., and Sego, D.C. 2009. Measurement of wind-induced pressure gradients in a waste rock pile. *Vadose Zone J.* 8(4):953–962. doi:10.2136/vzj2009.0002.

Anderson, K.F.E., Wall, F., Rollinson, G.K., and Moon, C.J. 2014. Quantitative mineralogical and chemical assessment of the Nkout iron ore deposit, Southern Cameroon. *Ore Geol. Rev.* 62:25–39.

Anderson, M.G., and Bates, P.D. 2001. *Model Validation: Perspectives in Hydrological Science*. New York: Wiley and Sons.

Anderson, P.W. 1963. *Variations in the Chemical Character of the Susquehanna River at Harrisburg, Pennsylvania*. U.S. Geological Survey Water-Supply Paper 1779-B. Washington, DC: U.S. Government Printing Office.

Antenucci, J., and Imerito, A. 2001. *The CWR Dynamic Reservoir Simulation Model DYRESM: Science Manual*. Report WP 1573 JA. Perth, Australia: Centre for Water Research, University of Western Australia.

APHA, AWWA, and WEF (American Public Health Association, American Water Works Association, and Water Environment Federation). 2012. *Standard Methods for the Examination of Water and Wastewater*, 22nd ed. Edited by E.W. Rice, R.B. Baird, A.D. Eaton, and L.S. Clesceri. Washington, DC: APHA.

Appelo, C.A.J., and Postma, D. 2007. *Geochemistry, Groundwater and Pollution*. Leiden, The Netherlands: A.A. Balkema.

Appelo, C.A.J., and Rolle, M. 2010. PHT3D: A reactive multicomponent transport model for reactive porous media. *Ground Water* 48:627–632.

Aquaveo LLC. 2016. Groundwater Modeling System, Surface-Water Modeling System, and Watershed Modeling System software products. Provo, UT: Aquaveo.

Arnold, J.G., Kiniry, J.R., Srinivasan, R., Williams, J.R., Haney, E.B., and Neitsch, S.L. 2012. *Soil and Water Assessment Tool: Input/Output Documentation, Version 2012*. Texas Water Resources Institute TR-439. http://swat.tamu.edu/media/69296/SWAT-IO-Documentation-2012.pdf (accessed July 23, 2013).

Arya, L.M., and Paris, J.F. 1981. A physico-empirical model to predict the soil moisture-characteristic from particle size distribution and bulk density data. *Soil Sci. Soc. Am. J.* 45:1023–1030.

Aryafar, A., Ardejani, F.D., Singh, R., and Shokri, B.J. 2007. Prediction of groundwater inflow and height of the seepage face in a deep open pit mine using numerical finite element model and analytical solutions. In *Water in Mining Environments: Proceedings IMWA Symposium 2007*. Edited by R. Cidu and F. Frau. Cagliari, Italy: Universita degli Studi di Cagliari.

Aubertin, M., Ricard, J.F., and Chapuis, R.P. 1998. A predictive model for the water retention curve: Application to tailings from hard-rock mines. *Can. Geotech. J.* 35(1):55–69.

Bain, J.G., Mayer, K.U., Molson, J.W.H., Blowes, D.W., Frind, E.O., Kahnt, R., and Jenk, U. 2001. Modelling the closure-related geochemical evolution of groundwater at a former uranium mine. *J. Contam. Hydrol.* 52(1):109–135.

Baker, B.J., and Banfield, J.F. 2003. Microbial communities in acid mine drainage. *FEMS Microbiol. Ecol.* 44(2):139–152.

Ball, J.W., and Nordstrom, D.K. 1985. Major and trace-element analyses of acid mine waters in the Leviathan mine drainage basin, California/Nevada—October, 1981 to October, 1982. Water-Resources Investigations Report 85-4169. Denver, CO: U.S. Geological Survey.

Ball, J.W., and Nordstrom, D.K. 1989. *Final Revised Analyses of Trace Elements from Acid Mine Waters in the Leviathan Mine Drainage Basin, California and Nevada; October 1981 to October 1982.* Water-Resources Investigations Report 89-4138. Denver, CO: U.S. Geological Survey.

Ball, J.W., and Nordstrom, D.K. 1991. *User's Manual for WATEQ4F, with Revised Thermodynamic Data Base and Test Cases for Calculating Speciation of Major, Trace, and Redox Elements in Natural Waters.* Open-File Report 91-183. Denver, CO: U.S. Geological Survey.

Ball, J.W., Runkel, R.L., and Nordstrom, D.K. 2004. Evaluating remedial alternatives for the Alamosa River and Wightman Fork, near the Summitville Mine, Colorado: Application of a reactive-transport model to low- and high-flow simulations. In *Environmental Sciences and Environmental Computing*, Vol. II. Edited by P. Zanetti. Fremont, CA: EnviroComp Institute. pp. 1–54.

Banks, D., Parnachev, V.P., Frengstad, B., Holden, W., Vedernikov, A.A., and Karnachuk, O.V. 2002. Alkaline mine drainage from metal sulphide and coal mines: Examples from Svalbard and Siberia. In *Mine Water Hydrogeology and Geochemistry*. Special Publication 198. Edited by P.L. Younger and N.S. Robins. London: The Geological Society. pp. 287–296.

Berner, E.K., and Berner, R.A. 1996. *Global Environment: Water, Air, and Geochemical Cycles*. Upper Saddle River, NJ: Prentice-Hall.

Berner, R.A. 1963. Electrode studies of hydrogen sulfide in marine sediments. *Geochim. Cosmochim. Acta* 27(6):563–575.

Bethke, C.M. 2007. *The Geochemist's Workbench® Release 7.0* (four volumes). Hydrogeology Program. Urbana, IL: University of Illinois.

Bethke, C.M. 2008. *Geochemical and Biogeochemical Reaction Modeling*, 2nd ed. Cambridge: Cambridge University Press.

Bethke, C.M., and Yeakel, S. 2014. *The Geochemist's Workbench User's Guides*, Version 10.0. Champaign, IL: Aqueous Solutions LLC.

Bigham, J.M., and Nordstrom, D.K. 2000. Iron and aluminum hydroxysulfates from acid sulfate waters. In *Sulfate Minerals—Crystallography, Geochemistry, and Environmental Significance*. Edited by C.N. Alpers, J.L. Jambor, and D.K. Nordstrom. Reviews in Mineralogy and Geochemistry, Vol. 40. Washington, DC: Mineralogical Society of America. pp. 351–403.

Bird, R.B., Stewart, W.E., and Lightfoot, E.M. 1960. *Transport Phenomena*. New York: John Wiley and Sons.

Blowes, D.W., Cherry, J.A., Reardon, E.J., and Jambor, J.L. 1991. The formation and potential importance of cemented layers in inactive sulfide mine tailings. *Geochim. Cosmochim. Acta* 55(4):965–978.

Blowes, D.W., Ptacek, C.J., Jambor, J.L., Weisener, C.G., Paktunc, D., Gould, W.D., and Johnson, D.B. 2014. The geochemistry of acid mine drainage. In *Environmental Geochemistry*. Edited by B. Sherwood Lollar. Treatise on Geochemistry, 2nd ed., Vol. 11, H.D. Holland and K.K. Turekian (editors in chief). Oxford: Elsevier Science. pp. 131–190.

Bredehoeft, J.D. 2003. From models to performance assessment: The conceptualization problem. *Ground Water* 41(5):571–577.

Bredehoeft, J.D. 2005. The conceptualization model problem—surprise. *Hydrogeol. J.* 13(1):37–46.

Brookins, D.G. 1988. *Eh-pH Diagrams for Geochemistry*. Berlin: Springer-Verlag.

Broshears, R.E., Runkel, R.L., Kimball, B.A., McKnight, D.M., and Bencala, K.E. 1996. Reactive solute transport in an acidic stream: Experimental pH increase and simulation of controls on pH, aluminum, and iron. *Environ. Sci. Technol.* 30(10):3016–3024.

Brown, J.G. 1996. Movement of metal contaminants in ground water in the Pinal Creek Basin, Arizona: Model assessment and simulation of reactive transport. Master of Science thesis, University of Arizona, Tucson.

Brown, J.G., and Glynn, P.D. 2003. Kinetic dissolution of carbonates and Mn oxides in acidic water: Measurement of in situ rates and reactive transport modeling. *Appl. Geochem.* 18(8):1225–1239.

Brown, J.G., Bassett, R., and Glynn, P.D. 1998. Analysis and simulation of reactive transport of metal contaminants in ground water in Pinal Creek Basin, Arizona. *J. Hydrol.* 209:225–250.

Brown, J.G., Bassett, R., and Glynn, P.D. 2000. Reactive transport of metal contaminants in alluvium—Model comparison and column simulation. *Appl. Geochem.* 15:35–49.

Brown, L.C., and Barnwell, T.O. 1987. *The Enhanced Stream Water Quality Models QUAL2E and QUAL2E-UNCAS*. EPA/600/3-87-007. Athens, GA: U.S. Environmental Protection Agency.

Brown, P.L., Curti, E., and Grambow, B. 2005. *Chemical Thermodynamics of Zirconium*. Edited by F. Mompean, J. Perrone, and M. Illemassene. Amsterdam: Elsevier.

Bruno, J., Bosbach, D., Kulik, D., and Navrotsky, A. 2007. *Chemical Thermodynamics of Solid Solutions of Interest in Radioactive Waste Management*. Edited by F.J. Mompean, M. Illemassene, and J. Perrone. Amsterdam: Elsevier.

Burns, D.A., Plummer, L.N., McDonnell, J.J., Busenberg, E., Casile, G.C., Kendall, C., Hooper, R.P., Freer, J.E., Peters, N.E., Beven, K., and Schlosser, P. 2003. The geochemical evolution of riparian ground water in a forested piedmont catchment. *Ground Water* 41(7):913–925.

Busenberg, E., and Plummer, L.N. 1982. The kinetics of dissolution of dolomite in CO_2-water systems at 1.5 to 65°C and 0.0 to 1.0 atmosphere pCO_2. *Am. J. Sci.* 282:45–78.

Butcher, A.R., Helms, T.A., Gottlieb, P., Bateman, R., Ellis, S., and Johnson, N.W. 2000. Advances in the quantification of gold deportment by QemSCAN. In *Seventh Mill Operators' Conference*. Kalgoorlie, Western Australia: Australasian Institute of Mining and Metallurgy. pp. 267–271.

Caley, E.R., and Richards, J.F.C. 1956. *Theophrastus on Stones*. Columbus, OH: Ohio State University.

Campbell, K.M., Nordstrom, D.K., and Hay, M.B. 2012. Kinetic modeling of microbial Fe(II) oxidation, Fe(III) hydrolysis, and As(III) oxidation in acid waters. In *International Congress on Arsenic in the Environment, Cairns, Australia, July 22–27, 2012*. Leiden: CRC Press. pp. 461–162.

Campbell, K.M., Alpers, C.N., Nordstrom, D.K., Blum, A.E., and Williams, A. 2013. Characterization and remediation of iron(III) oxide-rich scale in a pipeline carrying acid mine drainage at Iron Mountain mine, California, USA. In *International Mine Water Association Proceedings*. Golden, CO: IMWA. pp. 287–293.

Campbell, K.M., Gallegos, T.J., and Landa, E.R. 2015. Biogeochemical aspects of uranium mineralization, mining, milling, and remediation. *Appl. Geochem.* 57:206–235.

Castendyk, D. 2009. Chapter 9, Predictive modeling of the physical limnology of future pit lakes. In *Mine Pit Lakes: Characteristics, Predictive Modeling, and Sustainability*. Management Technologies for Metal Mining Influenced Water, Vol. 3. Edited by D. Castendyk and L.E. Eary. Littleton, CO: SME.

Castendyk, D.N., and Eary, L.E., eds. 2009. *Mine Pit Lakes: Characteristics, Predictive Modeling, and Sustainability*. Management Technologies for Metal Mining Influenced Water, Vol. 3. Littleton, CO: SME.

Chandra, A.P., and Gerson, A.R. 2010. The mechanisms of pyrite oxidation and leaching: A fundamental perspective. *Surf. Sci. Reports* 65(9):293–315.

Chapra, S.C., Pelletier, G.J., and Tao, H. 2008. *QUAL2K: A Modeling Framework for Simulating River and Stream Water Quality, Version 2.11: Documentation and Users Manual*. Medford, MA: Civil and Environmental Engineering Department, Tufts University.

Chase, M.W., Jr., ed. 1998. NIST-JANAF Thermochemical Tables, 4th ed. *J. Phys. Chem. Ref. Data*, Monograph 9, Parts I and II.

Chen, C.-J., and Jiang, W.-T. 2012. Influence of waterfall aeration and seasonal temperature variation on the iron and arsenic attenuation rates in an acid mine drainage system. *Appl. Geochem.* 27(10):1966–1978.

Cheng, H.P., and Yeh, G.T. 1998. Development of a three-dimensional model of subsurface flow, heat transfer, and reactive chemical transport: 3DHYDROGEOCHEM. *J. Contam. Hydrol.* 34(1-2):47–83.

Clement, T.P., and Johnson, C.D. 2012. RT3D: Reactive transport in 3-dimensions. In *Groundwater Reactive Transport Models*. Edited by F. Zhang, G.T. Yeh, and J.P. Parker. Sharjah, UAE: Bentham Science Publishers. pp. 96–111.

Cole, T.M., and S.A. Wells. 2015. *CE-QUAL-W2: A Two-Dimensional, Laterally Averaged, Hydrodynamic and Water Quality Model, Version 3.72 User's Manual*. Portland, OR: Department of Civil and Environmental Engineering, Portland State University.

Cornell, R.M., and Schwertmann, U. 2006. *The Iron Oxides: Structure, Properties, Reactions, Occurrences and Uses*, 2nd ed. Berlin: Wiley-VCH.

Cox, J.D., Wagman, D.D., and Medvedev, V.A. 1989. *CODATA Key Values for Thermodynamics*. New York: Hemisphere Publishing.

Craig, J. 2009. VisualAEM, graphical user interface for modeling, Version 1.04. Waterloo, ON: University of Waterloo. www.civil.uwaterloo.ca/jrcraig/visualaem/Main.html (accessed April 19, 2017).

Crock, J.G., Arbogast, B.F., and Lamothe, P.J. 1999. Laboratory methods for the analysis of environmental samples. In *The Environmental Geochemistry of Mineral Deposits*. Edited by G.S. Plumlee and M.J. Logsdon. Littleton, CO: Society of Economic Geologists. pp. 265–288.

Curtis, G.P., Davis, J.A., and Naftz, D.L. 2006. Simulation of reactive transport of uranium(VI) in groundwater with variable chemical conditions. *Water Resour. Res.* 42:W04404, 1–15.

Darcy, H. 1856. *Les fontaines publiques de la ville de Dijon* [The Public Fountains of the City of Dijon]. Paris: Victor Dalmont.

Dausman, A.M., Doherty, J., Langevin, C.D., and Dixon, J. 2010. Hypothesis testing of buoyant plume migration using a highly parameterized variable-density groundwater model. *Hydrolgeol. J.* 18:147–160.

Davis, G.B., and Ritchie, A.I.M. 1986. A model of oxidation in pyritic mine wastes: Part 1, Equations and approximate solution. *Appl. Math. Modeling* 10:314–322.

Deng, Y., Nordstrom, D.K., and McCleskey, R.B. 2011. Fluoride geochemistry of thermal waters in Yellowstone National Park: I. Aqueous fluoride speciation. *Geochim. Cosmochim. Acta* 75(16):4476–4489.

DHI Group. 2012. *MIKE SHE User Manual; Volume 1: User Guide*. http://dssplanning.dhigroup.com/links/MIKE_SHE_Printed_V1.pdf (accessed April 26, 2017).

DHI-WASY. 2012. *FEFLOW 6.1 User's Manual*. Berlin: DHI-WASY GmbH. www.feflow.info/ (accessed July 29, 2013).

Doherty, J. 2010. *PEST, Model-Independent Parameter Estimation User Manual*, 5th ed. Brisbane: Watermark Numerical Computing. www.pesthomepage.org/Home.php (accessed July 30, 2013).

Doherty, J.E., and Hunt, R.J. 2010. *Approaches to Highly Parameterized Inversion: A Guide to Using PEST for Groundwater-Model Calibration*. Scientific Investigations Report 2010-5169. Reston, VA: U.S. Geological Survey.

Dold, B., and Spangenberg, J.E. 2005. Sulfur speciation and stable isotope trends of water-soluble in mine tailings profiles. *Environ. Sci. Technol.* 39(15):5650–5656.

Domenico, P.A., and Schwartz, F.W. 1990. *Physical and Chemical Hydrogeology*. New York: John Wiley and Sons.

Duan, Q., Gupta, H.V., Sorooshian, S., Rousseau, A.N., and Turcotte, R., eds. 2003. *Calibration of Watershed Models*. Water Science and Application 6. Washington, DC: American Geophysical Union.

Dzomback, D.A., and Morel, F.M.M. 1990. *Surface Complexation Modeling: Hydrous Ferric Oxide*. New York: John Wiley and Sons.

Eary, T. 2007. Linking GoldSim with the PHREEQC geochemical model with a dynamic link library element. GoldSim 2007 User Conference, San Francisco. www.resolutionmineeis.us/sites/default/files/references/eary-2007.pdf and www.goldsim.com/Web/Solutions/EnvironmentalSystems/ (accessed May 25, 2017).

Engesgaard, P., and Christensen, T.H. 1988. A review of chemical solute transport models. *Nord. Hydrol.* 19:183–216.

ESI (Environmental Simulations Incorporated). 2017. Groundwater Vistas software, Version 6.88. Leesport, PA: ESI. http://groundwatermodels.com/ (accessed April 29, 2017).

Evans, J.R. 1966. Leviathan mine. In *Guidebook Along the East-Central Front of the Sierra Nevada*. Edited by J.R. Evans. Sacramento: Geological Society of Sacramento. pp. 63–75.

Fala, O., Aubertin, M., Molsin, J., Bussiere, B., Wilson, G.W., Chapuis, R., and Martin, V. 2003. Numerical modelling of unsaturated flow in uniform and heterogeneous waste rock piles. In *Proceedings of the Sixth International Conference on Acid Rock Drainage*. Carlton South, VIC: Australasian Institute of Mining and Metallurgy. pp. 895–902.

Fang, Y., Yabusaki, S.B., Morrison, S.J., Amonette, J.P., and Long, P.E. 2009. Multicomponent reactive transport modeling of uranium bioremediation field experiments. *Geochim. Cosmochim. Acta* 73(20):6029–6051.

Fayer, M.J., and Jones, T.L. 1990. *UNSAT-H Version 2.0: Unsaturated Soil Water and Heat Flow Model.* PNL-6779. Richland, WA: Pacific Northwest Laboratory.

Fenter, P.A., Rivers, M.L., Sturchio, N.C., Sutton, S.R., eds. 2002. *Applications of Synchrotron Radiation in Low-Temperature Geochemistry and Environmental Science.* Reviews in Mineralogy and Geochemistry, Vol. 49. Washington, DC: Geochemical Society and Mineralogical Society of America.

Fetter, C.W. 1993. *Contaminant Hydrogeology.* Upper Saddle River, NJ: Prentice-Hall.

Ficklin, W.H., Love, P.K., and Briggs, P.K. 1991. *Analytical Results for Total and Partial Metal Extractions in Aquifer Material, Pinal Creek, Globe, Arizona.* Open-File Report 91-111. Denver, CO: U.S. Geological Survey.

Fishel, D.K. 1984. *Water-Quality and Chemical Loads of the Susquehanna River at Harrisburg, Pennsylvania, April 1980 to March 1981.* Water-Resources Investigations Report 83-4164. Denver, CO: U.S. Geological Survey.

Fishman, M.J., and Friedman, L.C., eds. 1989. *Methods for Determination of Inorganic Substances in Water and Fluvial Sediments.* Techniques of Water-Resources Investigations 05-A1. Denver, CO: U.S. Geological Survey. http://nwql.usgs.gov/rpt.shtml?TWRI-5-A1

Fitts, C.R. 2010. Modeling aquifer systems with analytic elements and subdomains. *Water Resour. Res.* 46(7):WO7521.

Fitts Geosolutions. 2017. AnAqSim, Analytic Aquifer Simulator, Version 2017-1. Scarborough, ME: Fitts Geosolutions. www.fittsgeosolutions.com/ (accessed April 19, 2017).

Ford, A. 1999. *Modeling the Environment.* Washington, DC: Island Press.

Fredlund, D.G., Rahardjo, H., and Fredlund, M.D. 2012. *Unsaturated Soil Mechanics in Engineering Practice.* Hoboken, NJ: John Wiley and Sons.

Fredlund, M.D. 2004. *SoilVision, A Knowledge-Based Soils Database User's Manual.* ED-4A. Saskatoon, SK: SoilVision Systems. www.soilvision.com/subdomains/soildatabase.com/index.shtml (accessed July 29, 2013).

Fredlund, M.D. 2012. SVOffice 2009: 1D/2D/3D Finite Element Modeling Software Overview. Saskatoon, SK: SoilVision Systems. www.soilvision.com/subdomains/svflux.com/index.shtml (accessed July 29, 2013).

Freeze, R.A., and Cherry, J.A. 1979. *Groundwater.* Englewood Cliffs, NJ: Prentice-Hall.

Freeze, R.A., and Cherry, J.A. 1984. *Groundwater.* New York: Prentice-Hall.

Frind, E.O., and Molson, J.W. 1994. Modelling of mill tailings impoundments. In *Environmental Geochemistry of Sulfide Mine-Wastes.* Edited by J.L. Jambor and D.W. Blowes. Short Course Handbook, Vol. 22. Nepean, ON: Mineralogical Association of Canada. pp. 19–58.

Frippiat, C.C., and Holeyman, A.E. 2008. A comparative review of upscaling methods for solute transport in heterogeneous porous media. *J. Hydrol.* 362(1):150–176. doi:10.1016/j.jhydrol.2008.08.015

Furman, A. 2008. Modeling coupled surface-subsurface processes: A review. *Vadose Zone J.* 7(2):741–756.

Gammons, C. 2009. Sampling and monitoring of pit lakes. In *Mine Pit Lakes: Characteristics, Predictive Modeling, and Sustainability.* Management Technologies for Metal Mining Influenced Water, Vol. 3. Edited by D.M. Castendyk and L.E. Eary. Littleton, CO: SME. pp. 77–90.

Gamsjäger, H., Bugajski, J., Gajda, T., Lemire, R., and Preis, W. 2005. *Chemical Thermodynamics of Nickel.* Edited by F.J. Mompean, M. Illemassene, and J. Perrone. Amsterdam: Elsevier.

Gamsjäger, H., Gajda, T., Sangster, J., Saxena, S.K., and Voigt, W. 2012. *Chemical Thermodynamics of Tin.* Edited by J. Perrone. Amsterdam: Elsevier.

Garrels, R.M., and Christ, C.L. 1965. *Solutions, Minerals, and Equilibria.* New York: Harper's Geosciences.

Garvin, D., Parker, V.B., and White, H.J. 1987. *CODATA Thermodynamic Tables—Selections for Some Compounds of Calcium and Related Mixtures: A Prototype Set of Tables.* Washington, DC: Hemisphere Publishing.

Gelhar, L.W., Welty, C., and Rehfeldt, K.W. 1992. A critical review of data on field-scale dispersion in aquifers. *Water Resour. Res.* 28(7):1955–1974.

Geo-Slope International. 2012a. SEEP/W 2012 Groundwater Seepage Analysis. Calgary, AB: Geo-Slope International. www.geo-slope.com/products/seepw.aspx (accessed July 29, 2013).

Geo-Slope International. 2012b. VADOSE/W 2012 Vadose Zone and Soil Cover Analysis. Calgary, AB: Geo-Slope International. www.geo-slope.com/products/vadosew.aspx (accessed July 29, 2013).

Gerke, H.H., Frind, E.O., and Molson, J.W. 1998. Modelling the effect of chemical heterogeneity on acidification and solute leaching in overburden mine spoils. *J. Hydrol.* 209:166–185.

Geroni, J.N., and Sapsford, D.J. 2011. Kinetics of iron (II) oxidation determined in the field. *Appl. Geochem.* 26(8):1452–1457.

Glynn, P., and Brown, J. 1996. Reactive transport modeling of acidic metal-contaminated ground water at a site with sparse spatial information. In *Reactive Transport in Porous Media*. Edited by P.C. Lichtner, C.I. Steefel, and E.H. Oelkers. *Rev. Mineral.* 34:377–438.

Glynn, P., and Brown, J. 2012. Integrating field observations and inverse and forward modeling: Application at a site with acidic, heavy-metal-contaminated groundwater. In *Geochemical Modeling of Groundwater, Vadose and Geothermal Systems*. Edited by J. Bundschuh and M. Zilberbrand. Boca Raton, FL: CRC Press/Balkema. pp. 181–233.

Golder Associates Inc. 2011. Buckhorn Mine, updated groundwater flow model and mine inflow estimates. Submitted to Crown Resources/Kinross Kettle River Operations. Redmond, WA: Golder Associates.

Golder Associates Inc. 2012. FracMan 7.4 software. Redmond, WA: Golder Associates. http://fracman.golder.com/ (accessed July 30, 2013).

Golder Associates Inc. 2014. *Hydrologic Data Report—Water Year 2013*. Submitted to Crown Resources Corporation. Project No. 043 1077 129. Redmond, WA: Golder Associates.

Greenwood, H.J. 1989. On models and modeling. *Can. Mineral.* 27:1–14.

Grenthe, I., Fuger, J., Konings, R.J.M., Lemire, R.J., Muller, A.B., Nguyen-Trung, C., and Wanner, H. 1992. *Chemical Thermodynamics of Uranium*. Edited by H. Wanner and I. Forest. Amsterdam: Elsevier.

Guillamont, R., Fanghänel, T., Fuger, J., Grenthe, I., Neck, V., Palmer, D.A., and Rand, M.H. 2003. *Update on the Chemical Thermodynamics of Uranium, Neptunium, Plutonium, Americium, and Technetium* (and CD). Edited by F.J. Mompean, M. Illemassene, C. Domenech-Orti, and K. Ben Said. Amsterdam: Elsevier.

Gusek, J.J., and Figueroa, L.A., eds. 2009. *Mitigation of Metal Mining Influenced Water*. Management Technologies for Metal Mining Influenced Water, Vol. 2. Littleton, CO: SME.

Haitjema, H.M. 1995. *Analytic Element Modeling of Groundwater Flow*. San Diego: Academic Press.

Haitjema Software. 2016. GFLOW software, Version 2.2.2. www.haitjema.com (accessed April 19, 2017).

Hall, J.W., Middlemis, H., Waters, P.J., and Rozlapa, K.L. 2006. Integration of groundwater modelling with mine planning to optimize mine closure plans—The Marillana Creed (Yandi) mine story. In *Mine Closure 2006 Proceedings*. Edited by A. Fourie and M. Tibbett. Perth: Australian Centre for Geomechanics.

Hammond, G.E., Lichtner, P.C., Lu, C., and Mills, R.T. 2012. PFLOTRAN: Reactive flow and transport code for use on laptops to leadership-class supercomputers. In *Groundwater Reactive Transport Models*. Edited by F. Zhang, G.T. Yeh, and J.P. Parker. Sharjah, UAE: Bentham Science Publishers. pp. 141–159.

Hao, Y., Sun, Y., and Nitao, J.J. 2012. Overview of NUFT: A versatile numerical model for simulating flow and reactive transport in porous media. In *Groundwater Reactive Transport Models*. Edited by F. Zhang, G.T. Yeh, and J.P. Parker. Sharjah, UAE: Bentham Science Publishers. pp. 212–239.

Harbaugh, A.W. 2005. *MODFLOW-2005, The U.S. Geological Survey Modular Ground-Water Harbaugh Model—The Ground-Water Flow Process*. Techniques and Methods 6-A16. Reston, VA: U.S. Geological Survey. http://water.usgs.gov/nrp/gwsoftware/modflow.html.

Helgeson, H.C., Kirkham, D.H. and Flowers, G.C. 1981. Theoretical prediction of the thermodynamic behavior of aqueous electrolytes at high pressures and temperatures. IV. Calculation of activity coefficients and relative partial molal properties to 600°C and 5 kb. *Am. J. Sci.* 281:1249–1516.

Hem, J.D. 1972. Graphical methods for representing form and stability of aqueous metal ions. *Chem. Geol.* 9:119–132.

Hem, J.D. 1992. *Study and Interpretation of the Chemical Characteristics of Natural Water*, 3rd ed. Water-Supply Paper 2254. Alexandria, VA: U.S. Geological Survey.

Hem, J.D., and Cropper, W.H. 1959. *Survey of Ferrous-Ferric Chemical Equilibria and Redox Potentials*. Water-Supply Paper 1459-A. Washington, DC: U.S. Geological Survey.

HGL (HydroGeoLogic Inc.). 2012. MODFLOW-SURFACT, A Code for Analyzing Subsurface Systems. www.hglsoftware.com/Modflow.cfm (accessed July 29, 2013).

Hill, M.C., and Tiedeman, C.R. 2007. *Effective Groundwater Model Calibration: With Analysis of Data, Sensitivities, Predictions and Uncertainty*. New York: John Wiley and Sons.

Hillel, D. 1980. *Fundamentals of Soil Physics*. Orlando, FL: Academic Press.

Hodges, J.S. 1991. Six (or so) things you can do with a bad model. *Oper. Res.* 19:355–365.

Holland, T.J.B., and Powell, R. 2011. An improved and extended internally consistent thermodynamic dataset for phases of petrological interest, involving a new equation of state for solids. *J. Metamorph. Geol.* 29(3):333–383.

Hornberger, G.M., Raffensperger, J.P., Wiberg, P.L., and Eshleman, K.N. 1998. *Elements of Physical Hydrology*. Baltimore, MD: Johns Hopkins University Press.

Hostetler, J.D. 1984. Electrode electrons, aqueous electrons, and redox potentials in natural waters. *Am. J. Sci.* 284(6):734–759.

Hummel, W., Berner, U., Curti, E., Pearson, F.J., and Thoenen, T. 2002. *Nagra/PSI chemical Thermodynamic Data Base 01/01*. TDB Technical Report. Wettingen: Nagra.

Hummel, W., Anderegg, G., Puigdomenech, I., Rao, L. and Tochiyama, O. 2005. *Chemical Thermodynamics of Compounds and Complexes of U, Np, Pu, Am, Tc, Se, Ni and Zr with Selected Organic Ligands* (with CD). Edited by F.J. Mompean, J. Perrone, and M. Illemassene. Amsterdam: Elsevier.

Itasca Denver Inc. 2012. MINEDW software, Version 2.00.1. Lakewood, CO: Itasca Denver. www.itascadenver.com/minedw.html (accessed July 29, 2013).

Jambor, J.L. 2003. Mine-waste mineralogy and mineralogical perspectives of acid-base accounting. In *Environmental Aspects of Mine Wastes*. Edited by J.L. Jambor, D.W. Blowes, and A.I.M. Ritchie. Ottawa, ON: Mineralogical Association of Canada.

Jambor, J.L., Nordstrom, D.K., and Alpers, C.N. 2000. Metal sulfate salts from sulfide mineral oxidation. In *Sulfate Minerals—Crystallography, Geochemistry, and Environmental Significance*. Edited by C.N. Alpers, J.L. Jambor, and D.K. Nordstrom. Reviews in Mineralogy and Geochemistry, Vol. 40. Washington, DC: Mineralogical Society of America. pp. 305–350.

Javendel, I., Doughty, C., and Tsang, C.F. 1984. *Groundwater Transport: Handbook of Mathematical Models*. Water Resources Series Monograph 10. Washington, DC: American Geophysical Union.

Jenne, E.A. 1968. Controls on Mn, Fe, Co, Ni, Cu, and Zn concentrations in soils and water: The significant role of hydrous Mn and Fe oxides. In *Trace Inorganics in Water*. American Chemical Society Advances in Chemistry Series, Vol. 73. Washington, DC: American Chemical Society. pp. 337–387.

Johnson, J.W., Oelkers, E.H., and Helgeson, H.C. 1992. SUPCRT92: A software package for calculating the standard molal thermodynamic properties of minerals, gases, aqueous species, and reactions from 1 to 5,000 bars and 0° to 1,000°C. *Comp. Geosci.* 18(7):899–947.

Jørgensen, B.B., and Nelson, D.C. 2004. Sulfide oxidation in marine sediments: Geochemistry meets microbiology. In *Sulfur Biogeochemistry—Past and Present*. Edited by J. Amend, K.J. Edwards, and T.W. Lyons. Geological Society of America Special Papers, Vol. 379. Boulder, CO: Geological Society of America. pp. 63–81.

Karpenko, V., and Norris, J.A. 2002. Vitriol in the history of chemistry. *Chemicke Listy* 96:997–1005.

Keith, D.C., Runnells, D.D., Esposito, K.J., Chermak, J.A., Levy, D.B., Hannula, S.R., Watts, M., and Hall, L. 2001. Geochemical models of the impact of acidic groundwater and evaporative sulfate salts on Boulder Creek at Iron Mountain, California. *Appl. Geochem.* 16(7-8):947–961.

Kimball, B.A., Runkel, R.L., Walton-Day, K., and Bencala, K.E. 2002. Assessment of metal loads in watersheds affected by acid mine drainage by using tracer injection and synoptic sampling: Cement Creek, Colorado, USA. *Appl. Geochem.* 17(9):1183–1207.

Kimball, B.A., Nordstrom, D.K., Runkel, R.L., Vincent, K.R., and Verplanck, P.L. 2006. *Questa Baseline and Pre-Mining Ground-Water Quality Investigation. 23. Quantification of Mass Loading from Mined and Unmined Areas Along the Red River, New Mexico*. Scientific Investigations Report 2006-5004. Salt Lake City: U.S. Geological Survey.

Kirby, C.S., and Elder Brady, J.A. 1998. Field determinations of Fe^{2+} oxidation rates in acid mine drainage using a continuously-stirred reactor. *Appl. Geochem.* 13(4):509–520.

Kirby, C.S., Thomas, H.M., Southam, G., and Donald, R. 1999. Relative contributions of abiotic and biological factors in Fe(II) oxidation in mine drainage. *Appl. Geochem.* 14(4):511–530.

Kolditz, O., Bauer, S., Bilke, L., Böttcher, N., Delfs, J.O., Fischer, T., Görke, U.J., Kalbacher, T., Kosakowski, G., McDermott, C.I., Park, C.H., Radu, F., Rink, K., Shao, H., Shao, H.B., Sun, F., Sun, Y.Y., Singh, A.K., Taron, J., Walther, M., Wang, W., Watanabe, N., Wu, Y., Xie, M., Xu, W., and Zehner, B. 2012. OpenGeoSys: An open-source initiative for numerical simulation of thermos-hydro-mechanical/chemical (THM/C) processes in porous media. *Environ. Earth Sci.* 67:589–599.

Konikow, L.F., and Bredehoeft, J.D. 1992. Ground-water models cannot be validated. *Adv. Water Resour.* 15:75–83.

Kuang, J.-L., Huang, L.-N., Chen, L.-X., Hua, Z-S., Li, J.-J., Hu, M., Li, J.-T., and Shu, W.-S. 2013. Contemporary environmental variation determines microbial diversity patterns in acid mine drainage. *ISME J.* 7(5):1038–1050.

Kuipers, J.R., Maest, A., MacHardy, K.A., and Lawson, G. 2006. *Comparison of Predicted and Actual Water Quality at Hardrock Mines.* Washington, DC: Earthworks.

Kumpulainen, S., Raisanen, M.L., Von Der Kammer, F., and Hoffman, T. 2008. Ageing of synthetic and natural schwertmannites at pH 2–8. *Clay Miner.* 43(3):437–448.

Langmuir, D. 1971. The geochemistry of some carbonate ground waters in central Pennsylvania. *Geochim. Cosmochim. Acta* 35, 1023–1045.

Langmuir, D. 1997. *Aqueous Environmental Geochemistry.* Upper Saddle River, NJ: Prentice-Hall.

Lappala, E.G., Healy, R.W., and Weeks, E.P. 1987. *Documentation of Computer Program VS2D to Solve the Equations of Fluid Flow in Variably Saturated Porous Media.* Water-Resources Investigations Report 83-4099. Denver, CO: U.S. Geological Survey.

Lefebvre, R., Hockley, D., Smolensky, J., and Gélinas, P. 2001a. Multiphase transfer processes in waste rock piles producing acid mine drainage, 1. Conceptual model and system characterization. *J. Contam. Hydrol.* 52(1-4):137–164.

Lefebvre, R.D., Hockley, D., Smolensky, J., and Lamontagne, A. 2001b. Multiphase transfer processes in waste rock piles producing acid mine drainage 2: Applications of numerical simulation. *J. Contam. Hydrol.* 52(1-4):165–186.

Lemire, R.J., Fuger, J., Nitsche, H., Potter, P., Rand, M.H., Rydberg, J., Spahiu, K., Sullivan, J.C., Ullman, W.J., Vitorge, P., and Wanner, H. 2001. *Chemical Thermodynamics of Neptunium and Plutonium.* Amsterdam: Elsevier.

Lemire, R.J., Berner, U., Musikas, C., Palmer, D.A., Taylor, P., and Tochiyama, O. 2013. *Chemical Thermodynamics of Iron. Part 1.* Paris: Organisation for Economic Co-operation and Development Nuclear Energy Agency Data Bank.

Levenspiel, O. 1998. *Chemical Reaction Engineering*, 3rd ed. New York: John Wiley and Sons.

Lichtner, P.C. 1996. Continuum formulation of multicomponent-reactive transport. In *Reactive Transport in Porous Media.* Edited by P.C. Lichtner, C.I. Steefel, and E.H. Oelkers. Reviews in Mineralogy, Vol. 34. Washington, DC: Mineralogical Society of America.

Lichtner, P.C., Hammond, G.E., Lu, C., Karra, S., Bisht, G., Andre, B., Mills, R., and Kumar, J. 2015. *PFLOTRAN User Manual: A Massively Parallel Reactive Flow and Transport Model for Describing Surface and Subsurface Processes.* www.pflotran.org/docs/user_manual.pdf.

Lind, C.J., and Stollenwerk, K.G. 1994. Alteration of alluvium by acidic ground water resulting from copper mining of Pinal Creek, Arizona. In *U.S. Geological Survey Toxic Substances Hydrology Program; Proceedings of the Technical Meeting, Colorado Springs, Colorado, September 20–24, 1993.* Edited by D.W. Morganwalp and D.A. Aronson. Water-Resources Investigations Report 94-4015. Reston, VA: U.S. Geological Survey.

Lindberg, R.D., and Runnells, D.D. 1984. Groundwater redox reactions—An analysis of equilibrium state applied to Eh measurements and geochemical modeling. *Science* 225(4665):925–927.

Linklater, C.M., Bennett, J.W., and Edwards, N. 2006. Modelling of a waste rock dump design to control acid rock drainage at the Svartliden Gold Mine, northern Sweden. In *Proceedings of the Seventh International Conference on Acid Rock Drainage.* Lexington, KY: American Society of Mining and Reclamation. pp. 1079–1105. www.asmr.us/Portals/0/Documents/Conference-Proceedings/2006/1079-Linklater.pdf.

Ludington, S.A., Plumlee, G.S., Caine, J.S., Bove, D.J., Holloway, J.M., and Livo, K.E. 2005. *Questa Baseline and Pre-Mining Ground-Water Quality Investigation. 10. Geologic Influences on Ground and Surface Waters in the Lower Red River Watershed, New Mexico.* Scientific Investigations Report 2004-5245. Reston, VA: U.S. Geological Survey.

Maest, A.S., and Kuipers, J.R. 2005. *Predicting Water Quality at Hardrock Mines: Methods and Models, Uncertainties and State-of-the-Art.* Washington, DC: Earthworks.

Mangold, D.C., and Tsang, C.-F. 1991. A summary of subsurface hydrological and hydrochemical models. *Rev. Geophys.* 29(1):51–79.

Marinelli, F., and Niccoli, W.L. 2000. Simple analytical equations for estimating ground water inflow to a mine pit. *Ground Water* 38(2):311–314.

Markstrom, S.L., Niswonger, R.G., Regan, R.S., Prudic, D.E., and Barlow, P.M. 2008. *GSFLOW—Coupled Ground-Water and Surface-Water FLOW Model Based on the Integration of the Precipitation-Runoff Modeling System (PRMS) and the Modular Ground-Water Flow Model (MODFLOW-2005).* Techniques and Methods 6-D1. Reston, VA: U.S. Geological Survey. http://water.usgs.gov/nrp/gwsoftware/gsflow/gsflow.html

Martell, A.E., and Smith, R.M. 1974–1989. *Amino Acids*, Vol. 1; *Amines*, Vol. 2; *Other Organic Ligands*, Vol. 3; *Inorganic Compounds*, Vol. 4; *First Supplement*, Vol. 5; *Second Supplement*, Vol. 6. Critical Stability Constants Series. New York: Plenum.

Massmann, J.W. 1989. Applying groundwater flow models in vapor extraction system design. *J. Environ. Eng.* 115(1):129–149.

Massmann, J.W., and Farrier, D.F. 1992. Effects of atmospheric pressures on gas transport in the vadose zone. *Water Resour. Res.* 28(3):777–791.

Mayer, K.U. 1999. A numerical model for the investigation of reactive transport in variably-saturated porous media. Doctoral thesis, Department of Earth Sciences, University of Waterloo, Ontario.

Mayer, K.U., Blowes, D.W., and Frind, E.O. 2000. Numerical modeling of acid mine drainage generation and subsequent reactive transport. In *ICARD 2000, Proceedings from the 5th International Conference on Acid Rock Drainage*, Vol. 1. Littleton, CO: SME. pp. 135–142.

Mayer, K.U., Frind, E.O., and Blowes, D.W. 2002. A numerical model for the investigation of reactive transport in variably saturated media using a generalized formulation for kinetically controlled reactions. *Water Resour. Res.* 38(9):13-1–13-21.

Mayer, K.U., Blowes, D.W., and Frind, E.O. 2003. Advances in reactive transport modeling of contaminant release and attenuation from mine waste deposits. In *Environmental Aspects of Mine Wastes*. Edited by J.L. Jambor, D.W. Blowes, and A.I.M. Ritchie. Ottawa, ON: Mineralogical Association of Canada. pp. 283–302.

Mayer, K.U., Amos, R.T., Molins, S., and Gerard, F. 2012. Chapter 8, Reactive transport modeling in variably saturated media with MIN3P: Basic model formulation and model enhancements. In *Groundwater Reactive Transport Models*. Edited by F. Zhang, G.T. Yeh, J.C. Parker, and X. Shi. Sharjah, UAE: Bentham Science Publishers. pp. 187–212.

McCleskey, R.B. 2013. New method for electrical conductivity temperature compensation. *Environ. Sci. Technol.* 47(17):9874–9881.

McCleskey, R.B., Nordstrom, D.K. and Naus, C.A. 2004. *Questa Baseline and Pre-Mining Ground-Water Quality Investigation. 16. Quality Assurance and Quality Control of Water Analyses*. Open-File Report 2004-1341. Boulder, CO: U.S. Geological Survey.

McCleskey, R.B., Nordstrom, D.K., Ryan, J.N., and Ball, J.W. 2012. A new method of calculating electrical conductivity with applications to natural waters. *Geochim. Cosmochim. Acta* 77:369–382.

McKnight, D.M., and Bencala, K.E. 1989. Reactive iron transport in an acidic mountain stream in Summit County, Colorado: A hydrologic perspective. *Geochim. Cosmochim. Acta* 53:2225–2234.

McLemore, V.T., ed. 2008. *Basics of Metal Mining Influenced Water*. Management Technologies for Metal Mining Influenced Water, Vol. 1. Littleton, CO: SME.

McLemore, V.T., Smith, K.S., and Russell, C.C., eds. 2014. *Sampling and Monitoring for the Mine Life Cycle*. Management Technologies for Metal Mining Influenced Water, Vol. 6. Englewood, CO: SME.

MEND (Mine Environment Neutral Drainage). 1995. *MINEWALL 2.0, Series of Four Reports: Literature Review, User's Guide, Application of MINEWALL to Three Minesites, and Programmer Notes and Source Code*, plus one diskette. Ottawa, ON: MEND.

Merkel, B.J., and Planer-Friedrich, B. 2008. *Groundwater Geochemistry: A Practical Guide to Modeling of Natural and Contaminated Aquatic Systems*, 2nd ed. Edited by D.K. Nordstrom. Berlin: Springer-Verlag.

Millero, F.J. 2001. *Physical Chemistry of Natural Waters*. New York: Wiley-Interscience.

Mockus, V. 1972. Section 4, Hydrology; Chapter 21, Design hydrographs. In *National Engineering Handbook*. NEH Notice 4-102. Washington, DC: Soil Conservation Service.

Molson, J.W., Fala, O., Aubertin, M., and Bussière, B. 2005. Numerical simulations of pyrite oxidation and acid mine drainage in unsaturated waste rock piles. *J. Contam. Hydrol.* 78(4):343–371.

Moore, J.N., and Luoma, S.N. 1990. Hazardous wastes from large-scale metal extraction. *Environ. Sci. Technol.* 24:1279–1285.

Mueller, M. 2013. PITLAKQ, the Pit Lake Hydrodynamic and Water Quality Model. www.pitlakq.com/ (accessed July 30, 2013).

Murphy, R., and Strongin, D.R. 2009. Surface reactivity of pyrite and related sulfides. *Surf. Sci. Reports* 64(1):1–45.

Narasimhan, T.N. 1995. Models and modeling of hydrogeologic processes. *Soil Sci. Soc. Am. J.* 59:300–306.

Narasimhan, T.N., White, A.F. and Tokunga, T. 1986. Groundwater contamination from an inactive uranium mill tailings piled, 2. Application of a dynamic mixing model. *Water Resour. Res.* 22:1820–1834.

Neuman, S.P. 1973. Saturated-unsaturated seepage by finite elements. *J. Hydraul. Div. ASCE* 99:2233–2250.

Newsome, L., Morris, K., and Lloyd, J.R. 2014. The biogeochemistry and bioremediation of uranium and other priority radionuclides. *Chem. Geol.* 363:164–184.

Nicholson, A., Davis, A., and Helgen, S. 2003. Elements influencing cost apportionment in the Pinal Creek Aquifer, Arizona, USA. Part II: Identification of geochemical controls on remediation time. *Environ. Forensics* 4:271–286.

Nimick, D.A., Gammons, C.H., and Parker, S.R. 2011. Diel biogeochemical processes and their effect on the aqueous chemistry of streams: A review. *Chem. Geol.* 283:3–17.

Nordstrom, D.K. 1977. Hydrogeochemistry and microbiology of an acid mine drainage system. Doctoral thesis, Stanford University, California.

Nordstrom, D.K. 1996. Trace metal speciation in natural waters: Computational vs. analytical. *Water Air Soil Pollut.* 90:257–267.

Nordstrom, D.K. 2008. *Questa Baseline and Pre-Mining Ground-Water Quality Investigation. 25. Summary of Results and Baseline and Pre-Mining Ground-Water Geochemistry, Red River Valley, Taos County, New Mexico, 2001–2005.* Professional Paper 1728. Reston, VA: U.S. Geological Survey.

Nordstrom, D.K. 2009. Acid rock drainage and climate change. *J. Geochem. Explor.* 100:97–104.

Nordstrom, D.K. 2011. Hydrogeochemical processes governing the origin, transport, and fate of major and trace elements from mine wastes and mineralized rock to surface waters. *Appl. Geochem.* 26:1777–1791.

Nordstrom, D.K. 2012. Models, validation, and applied geochemistry; issues in science, communication, and philosophy. *Appl. Geochem.* 27:1899–1919.

Nordstrom, D.K., and Alpers, C.N. 1999. Negative pH, efflorescent mineralogy, and consequences for environmental restoration at the Iron Mountain Superfund site, California. *Proc. Natl. Acad. Sci.* 96:3455–3462.

Nordstrom, D.K., and Ball, J.W. 1986. The geochemical behavior of aluminum in acidified surface waters. *Science* 232:54–56.

Nordstrom, D.K., and Campbell, K.M. 2014. Modeling low-temperature geochemical processes. In *Treatise of Geochemistry*. Edited by H.D. Holland and K.K. Turekian. Vol. 7, Surface and Ground Water, Weathering, and Soils. Edited by J.I. Drever. Amsterdam: Elsevier. pp. 27–68.

Nordstrom, D.K., and May, H.M. 1996. Aqueous equilibrium data for mononuclear aluminum species. In *The Environmental Chemistry of Aluminum*, 2nd ed. Edited by G. Sposito. Boca Raton, FL: CRC Press. pp. 39–80.

Nordstrom, D.K., and McCleskey, R.B. 2012. Improving the reliability of water analyses: Beyond QA/QC. Presented at 9th International Symposium on Environmental Geochemistry, Aveiro, Portugal, July 15–21, 2012. p. 168 (abstract).

Nordstrom, D.K., and Munoz, J.L. 1994. *Geochemical Thermodynamics*. Boston: Blackwell Scientific.

Nordstrom, D.K., Plummer, L.N., Wigley, T.M.L., Wolery, T.J., Ball, J.W., Jenne, E.A., Bassett, R.L., Crerar, D.A., Florence, T.M., Fritz, B., Hoffman, M., Holdren, G.R., Jr., Lafon, G.M., Mattigod, S.V., McDuff, R.E., Morel, F., Reddy, M.M., Sposito, G., and Thrailkill, J. 1979. A comparison of computerized chemical models for equilibrium calculations in aqueous systems. In *Chemical Modeling of Aqueous Systems*, Vol. 93. Edited by E.A. Jenne. Washington, DC: American Chemical Society. pp. 857–892.

Nordstrom, D.K., Ball, J.W., Donahoe, R.J., and Whittemore, D. 1989. Groundwater chemistry and water rock interactions at Stripa. *Geochim. Cosmochim. Acta* 53:1727–1740.

Nordstrom, D.K., Alpers, C.N., Ptacek, C.J., and Blowes, D.W. 2000. Negative pH and extremely acidic mine waters from Iron Mountain, California. *Environ. Sci. Technol.* 34(2):254–258.

Nordstrom, D.K., Wright, W.G., Mast, M.A., Bove, D.J., and Rye, R.O. 2007. Aqueous-sulfate stable isotopes—A study of mining-affected and undisturbed acidic drainage. In *Integrated Investigations of Environmental Effects of Historical Mining in the Animas River Watershed, San Juan County, Colorado.* Edited by S.E. Church, P. von Guerard, and S.E. Finger. Professional Paper 1651. Reston, VA: U.S. Geological Survey. pp. 387–416.

Nordstrom, D.K., McCleskey, R.B., and Ball, J.B. 2010. Challenges in the analysis and interpretation of acidic waters. In *International Mine Water Association Symposium 2010: Mine Water and Innovative Thinking*. Sydney, NS: Cape Breton University Press. pp. 43–46.

NRC (National Research Council). 2007. *Models in Environmental Regulatory Decision Making*. Washington, DC: National Academies Press.

Oelkers, E.H., Helgeson, H.C., Shock, E.L., Sverjensky, D.A., Johnson, J.W., and Pokrovskii, V.A. 1995. Summary of the apparent partial molal Gibbs free energies of formation of aqueous species, minerals, and gases at pressures 1 to 5000 bars and temperatures 25 to 1000°C. *J. Phys. Chem. Ref. Data* 24:1401–1506.

Ogata, A., and Banks, R.B. 1961. *A Solution of the Differential Equation of Longitudinal Dispersion in Porous Media*. Professional Paper 411-A. Washington, DC: U.S. Government Printing Office. http://pubs.er.usgs.gov/publication/pp411A.

Olin, Å., Noläng, B., Osadchii, E.G., Öhman, L.-O., and Rosé, E. 2005. *Chemical Thermodynamics of Selenium* (with CD). Edited by F.J. Mompean, J. Perrone, and M. Illemassene. Amsterdam: Elsevier.

Oreskes, N. 2000a. Why believe a computer? Models, measures, and meaning in the natural world. In *The Earth Around Us—Maintaining a Livable Planet*. Edited by J.S. Schneiderman. New York: W.H. Freeman. pp. 70–82.

Oreskes, N. 2000b. Why predict? Historical perspectives on prediction in earth science. In *Prediction—Science, Decision Making, and the Future of Nature*. Edited by D. Sarewitz, R.A. Pielke Jr., and R. Byerly Jr. Washington, DC: Island Press. pp. 23–40.

Oreskes, N., Shrader-Frechette, K., and Belitz, K. 1994. Verification, validation, and confirmation of numerical models in the earth sciences. *Science* 263:641–646.

Panday, S., Huyakorn, P.S., Therrien, R., and Nichols, R.L. 1993. Improved three-dimensional finite-element techniques for field simulation of variably saturated flow and transport. *J. Contam. Hydrol.* 12:3–33.

Parkhurst, D.L. 1997. Geochemical mole-balancing with uncertain data. *Water Resour. Res.* 33:1957–1970.

Parkhurst, D.L., and Appelo, C.A.J. 1999. *User's Guide to PHREEQC (Version 2)—A Computer Program for Speciation, Batch-Reaction, One-Dimensional Transport, and Inverse Geochemical Calculations*. Water-Resources Investigations Report 99-4259. Denver, CO: U.S. Geological Survey.

Parkhurst, D.L., and Appelo, C.A.J. 2013. *Description and Input and Examples for PHREEQC Version 3—A Computer Program for Speciation, Batch-Reaction, One-Dimensional Transport, and Inverse Geochemical Calculations*. Techniques and Methods 6-A43. Denver, CO: U.S. Geological Survey.

Parkhurst, D.L., and Plummer, L.N. 1993. Geochemical models. In *Regional Ground-Water Quality*. Edited by W.N. Alley. New York: Van Nostrand Reinhold. pp. 32–58.

Parkhurst, D.L., Kipp, K.L., and Charlton, S.R. 2010. *PHAST Version 2—A Program for Simulating Groundwater Flow, Solute Transport, and Multicomponent Geochemical Reactions*. Techniques and Methods 6-A35. Denver, CO: U.S. Geological Survey.

Pauling, L. 1964. *College Chemistry*, 3rd ed. San Francisco: W.H. Freeman.

Peratta, A., and Popov, V. 2006. A new scheme for numerical modelling of flow and transport processes in 3D fractured porous media. *Adv. Water Resour.* 29:42–61. doi:10.1016/j.advwatres.2005.05.004

Perkins, E.H., Nesbitt, H.W., Gunter, W.D., St-Arnaud, L.C., and Mycroft, J.R. 1995. *Critical Review of Geochemical Processes and Geochemical Models Adaptable for Prediction of Acidic Drainage from Waste Rock*. MEND Report 1.42.1. Ottawa, ON: Mine Environment Neutral Drainage Program.

Pilkey, O.H., and Pilkey-Jarvis, L. 2007. *Useless Arithmetic: Why Environmental Scientists Can't Predict the Future*. New York: Columbia University Press.

Pirrie, D., Butcher, A.R., Power, M.R., Gottlieb, P., and Miller, G.L. 2004. Rapid quantitative mineral and phase analysis using automated scanning electron microscopy (QemSCAN); potential applications in forensic geoscience. *Geol. Soc. Spec. Pub.* 232:123–136.

Pitzer, K.S. 1991. Ion interaction approach: Theory and data correlation. In *Activity Coefficients in Electrolyte Solutions*. Edited by K.S. Pitzer. Boca Raton, FL: CRC Press. pp. 75–154.

Plumlee, G., Lowers, H., Ludington, S., Koenig, A., and Briggs, P. 2005. *Questa Baseline and Pre-Mining Ground-Water Quality Investigation. 13. Mineral Microscopy and Chemistry of Mined and Unmined Porphyry Molybdenum Mineralization Along the Red River, New Mexico—Implications for Ground- and Surface Water Quality*. Open-File Report 2005–1442. Reston, VA: U.S. Geological Survey.

Plummer, L.N., and Back, W. 1980. The mass balance approach: application to interpreting the chemical evolution of hydrologic systems. *Am. J. Sci.* 280:130–142.

Plummer, L.N., Wigley, T.M.L., and Parkhurst, D.L. 1978. The kinetics of calcite dissolution in CO_2-water systems at 5 to 65°C and 0.0 to 1.0 atmosphere pCO_2. *Am. J. Sci.* 278:179–216.

Plummer, L.N., Parkhurst, D.L., and Thorstenson, D.C. 1983. Development of reaction models for ground-water systems. *Geochim. Cosmochim. Acta* 47:665–686.

Poeter, E., Hill, M.C., Banta, E.R., Mehl, S., and Christensen, S. 2008. *UCODE_2005 and Six Other Computer Codes for Universal Sensitivity Analysis, Calibration, and Uncertainty Evaluation*. Techniques and Methods 6-A11. Reston, VA: U.S. Geological Survey.

Popper, K. 1934. *The Logic of Scientific Discovery* [originally in German as Logik der Froschung]. Vienna: Verlag von Julius Springer.

Postma, D., and Appelo, C.A.J. 2000. Reduction of Mn oxides by ferrous iron in a flow system: Column experiment and reactive transport modeling. *Geochim. Cosmochim. Acta* 64:1237–1247.

Powell, K.J., Brown, P.L., Byrne, R.H., Gajda, T., Hefter, G., Sjöberg, S., and Wanner, H. 2005. Chemical speciation of environmentally significant heavy metals with inorganic ligands. Part 1: The Hg^{2+}–Cl^-, OH^-, CO_3^{2-}, SO_4^{2-}, and PO_4^{3-} aqueous systems. *Pure Appl. Chem.* 77(4):739–800.

Powell, K.J., Brown, P.L., Byrne, R.H., Gajda, T., Hefter, G., Sjöberg, S., and Wanner, H. 2007. Chemical speciation of environmentally significant heavy metals with inorganic ligands. Part 2: The Cu^{2+}–Cl^-, OH^-, CO_3^{2-}, SO_4^{2-}, and PO_4^{3-} aqueous systems. *Pure Appl. Chem.* 79(5):895–950.

Powell, K.J., Brown, P.L., Byrne, R.H., Gajda, T., Hefter, G., Leuz, A.-K., Sjöberg, S., and Wanner, H. 2009. Chemical speciation of environmentally significant heavy metals with inorganic ligands. Part 3: The Pb^{2+}–Cl^-, OH^-, CO_3^{2-}, SO_4^{2-}, and PO_4^{3-} aqueous systems. *Pure Appl. Chem.* 81(12):2425–2476.

Powell, K.J., Brown, P.L., Byrne, R.H., Gajda, T., Hefter, G., Leuz, A.-K., Sjöberg, S., and Wanner, H. 2011. Chemical speciation of environmentally significant heavy metals with inorganic ligands. Part 4: The Cd^{2+}–Cl^-, OH^-, CO_3^{2-}, SO_4^{2-}, and PO_4^{3-} aqueous systems. *Pure Appl. Chem.* 83(5):1163–1214.

Powell, K.J., Brown, P.L., Byrne, R.H., Gajda, T., Hefter, G., Leuz, A.-K., Sjöberg, S., and Wanner, H. 2013. Chemical speciation of environmentally significant heavy metals with inorganic ligands. Part 5: The Zn^{2+}–Cl^-, OH^-, CO_3^{2-}, SO_4^{2-}, and PO_4^{3-} aqueous systems. *Pure Appl. Chem.* 85(12):2249–2311.

Prommer, H., Barry, D.A., and Zheng, C. 2003. MODFLOW/MT3DMS based reactive multi-component transport model. *Ground Water* 42:247–257.

Pruess, K., Oldenburg, C., and Moridis, G. 2012. *TOUGH2 User's Guide, Version 2.* Report LBL-43134. Berkeley, CA: Lawrence Berkeley Laboratory.

Ptacek, C., and Blowes, D. 2000. Predicting sulfate-mineral solubility in concentrated waters. In *Sulfate Minerals: Crystallography, Geochemistry, and Environmental Significance.* Edited by C.N. Alpers, J.L. Jambor, and D.K. Nordstrom. Reviews in Mineralogy and Geochemistry, Vol. 40. Washington, DC: Mineralogical Society of America, Geochemical Society. pp. 513–540.

Rand, M., Fuger, J., Grenthe, I., Neck, V., and Rai, D. 2009. *Chemical Thermodynamics of Thorium* (with CD). Edited by M. Illemassene, J. Perrone, and F.J. Mompean. Amsterdam: Elsevier.

Rard, J.A., Rand, M.H., Anderegg, G., and Wanner, H. 1999. *Chemical Thermodynamics of Technetium.* Edited by M.C.A. Sandino and E. Östhols. Amsterdam: Elsevier.

Raudsepp, M., and Pani, E. 2003. Application of Rietveld analysis to environmental mineralogy. In *Environmental Aspects of Mine Wastes.* Edited by J.L. Jambor, D.W. Blowes, and A.I.M. Ritchie. Ottawa, ON: Mineralogical Association of Canada. pp. 165–180.

Richards, L.A. 1931. Capillary conduction of liquids in porous mediums. *J. Phys.* 1:318–333.

Rimstidt, J.D., and Vaughan, D.J. 2003. Pyrite oxidation: A state-of-the-art assessment of the reaction mechanism. *Geochim. Cosmochim. Acta* 67:873–880.

Robie, R.A., and Hemingway, B.S. 1995. *Thermodynamic Properties of Minerals and Related Substances at 298.15 K and 1 Bar (10^5 Pascals) Pressure and at Higher Temperatures.* Bulletin 2131. Reston, VA: U.S. Geological Survey.

Romano, C.G., Mayer, K.U., and Blowes, D.W. 2006. Reactive transport modeling of AMD release and attenuation at the Fault Lake Tailings Area, Falconbridge, Ontario. In *Conference Proceedings of 7th ICARD, March 26–30, St. Louis, MO.* Edited by R. Barnhisel. Lexington, KY: ASMR.

Rubin, J. 1983. Transport of reacting solutes in porous media: Relation between mathematical nature of problem formulation and chemical nature of reactions *Water Resour. Res.* 19:1231–1252.

Runkel, R.L. 1998. *One-Dimensional Transport with Inflow and Storage (OTIS): A Solute Transport Model for Streams and Rivers*. Water-Resources Investigations Report 98-4018. Denver, CO: U.S. Geological Survey.

Runkel, R.L. 2010. *One-Dimensional Transport with Equilibrium Chemistry (OTEQ): A Reactive Transport Model for Streams and Rivers*. Techniques and Methods 6-B6. Reston, VA: U.S. Geological Survey.

Runkel, R.L., and Kimball, B.A. 2002. Evaluating remedial alternatives for an acid mine drainage stream: Application of a reactive transport model. *Environ. Sci. Technol.* 36:1093–1101.

Runkel, R.L., McKnight, D.M., and Bencala, K.E. 1996. Reactive solute transport in streams 2. Simulation of a pH modification experiment. *Water Resour. Res.* 32:419–430.

Runkel, R.L., Crawford, C.G., and Cohn, T.A. 2004. *Load Estimator (LOADEST): A FORTRAN Program for Estimating Constituent Load in Streams and Rivers*. Techniques and Methods 4-A5. Reston, VA: U.S. Geological Survey.

Runkel, R.L., Kimball, B.A., Walton-Day, K., Verplanck, P.L., and Broshears, R.E. 2012. Evaluating remedial alternatives for an acid mine drainage stream: A model post audit. *Environ. Sci. Technol.* 46:340–347.

Runkel, R.L., Walton-Day, K., Kimball, B.A., Verplanck, B.A., Verplanck, P.L., and Nimick, D.A. 2013. Estimating instream constituent loads using replicate synoptic sampling, Peru Creek, Colorado. *J. Hydrol.* 499:26–41.

Saaltink, M.W., Carrera, J., and Ayora, C. 2001. On the behavior of approaches to simulate reactive transport, *J. Contam. Hydrol.* 48:213–235.

Saaltink, M.W., Batle, F., Ayora, C., Carrera, J., and Olivella, S. 2004. RETRASO, a code for modeling reactive transport in saturated and unsaturated porous media. *Geol. Acta* 2:235–251.

Sato, M. 1992. Persistency-field Eh-pH diagrams for sulfides and their application to supergene oxidation and enrichment of sulfide ore bodies. *Geochim. Cosmochim. Acta* 56:3133–3156.

Scharer, J.M., Nicholson, R.V., Halbert, B., and Snodgrass, W.J. 1994. A computer program to assess acid generation in pyritic tailings. In *Environmental Geochemistry of Sulfide Oxidation*. Symposium Series, Vol. 550. Washington, DC: American Chemical Society. pp. 132–152.

Scharffenberg, W.A., and Fleming, M.J. 2010. *Hydrologic Modeling System HEC-HMS: User's Manual (Version 3.5)*. Publication CPD-74A. Davis, CA: U.S. Army Corps of Engineers, Hydrologic Engineering Center. www.hec.usace.army.mil/software/hec-hms/documentation.aspx (accessed July 25, 2013).

Schippers, A. 2004. Biogeochemistry of metal sulfide oxidation in mining environments, sediments and soils. In *Sulfur Biogeochemistry—Past and Present*. Edited by J. Amend, K.J. Edwards, and T.W. Lyons. Special Papers 379. Boulder, CO: Geological Society of America. pp. 49–62.

Schippers, A., Breuker, A., Blazejak, A., Bosecker, K., Kock, D., and Wright, T.L. 2010. The biogeochemistry and microbiology of sulfidic mine waste and bioleaching dumps and heaps, ad noverl Fe(II)-oxidizing bacteria. *Hydrometallurgy* 104(3-4):342–350.

Schroeder, P.R., Dozier, T.S., Zappi, P.A., McEnroe, B.M., Sjostrom, J.W., and Peyton, R.L. 1994. *The Hydrologic Evaluation of Landfill Performance (HELP) Model: Engineering Documentation for Version 3*. EPA/600/R-94/168b. Washington, DC: U.S. Environmental Protection Agency.

Šejna, M., Šimůnek, J., and van Genuchten, M.Th. 2011. *The HYDRUS Software Package for Simulating Two- and Three-Dimensional Movement of Water, Heat, and Multiple Solutes in Variably-Saturated Media, User Manual, Version 2.0*. Prague, Czech Republic: PC Progress. www.pc-progress.com/en/Default.aspx?hydrus-3d (accessed July 29, 2013).

Shaff, J.E., Schultz, B.A., Craft, E.J., Clark, R.T., and Kochian, L.V. 2010. GEOCHEM-EZ: A chemical speciation program with greater power and flexibility. *Plant Soil* 330(1):207–214.

Shock, E.L., Sassani, D.C., Willis, M., and Sverjensky, D.A. 1997. Inorganic species in geologic fluids: Correlations among standard molal thermodynamic properties of aqueous ions and hydroxide complexes. *Geochim. Cosmochim. Acta* 61:907–950.

Sigg, L. 2000. *Redox Potential Measurements in Natural Waters: Significance, Concepts and Problems*. Edited by J. Schüring, H.D. Schulz, W.R. Fischer, J. Böttcher, and W.H.M. Duijnisveld. Berlin: Springer-Verlag. pp. 1–12.

Silva, R.J., Bidoglio, G., Rand, M.H., Robouch, P.B., Wanner, H., and Puigdomenech, I. 1995. *Chemical Thermodynamics of Americium*. Amsterdam: Elsevier.

Šimůnek, J., and Suarez, D.L. 1994. Two-dimensional transport model for variably saturated porous media with major ion chemistry. *Water Resour. Res.* 30:1115–1133.

Šimůnek, J., van Genuchten, M.Th., Šejna, M., Toride, N., and Leij, F.J. 1999. *The STANMOD Computer Software for Evaluating Solute Transport in Porous Media Using Analytical Solutions of Convection-Dispersion Equation* (Versions 1.0 and 2.0). IGWMC-TPS 71. Golden, CO: Colorado School of Mines, International Ground Water Modeling Center. www.pc-progress.com/Documents/Jirka/Stanmod.pdf.

Šimůnek, J., Jarvis, N.J., van Genuchten, M.Th., and Gärdenäs, A. 2003. Review and comparison of models for describing non-equilibrium and preferential flow and transport in the vadose zone. *J. Hydrol.* 272:14–35.

Šimůnek, J., Jacques, D., Langergraber, G., Bradford, S.A., Šejna, M., and van Genuchten, M.Th. 2013. Numerical modeling of contaminant transport using HYDRUS and its specialized modules. *J. Ind. Inst. Sci.* 93(2):265–284.

Šimůnek, J., Jarvis, N.J., van Genuchten, M.Th., and Šejna, M. 2016. Recent developments and applications of the HYDRUS computer software packages. *Vadose Zone J.* 15(8):1539–1663.

Song, X., and Leone, G. 2013. Modeling application in evaluating environmental impact due to phosphate mining activities. In *Reliable Mine Water Technology: Proceedings of the International Mine Water Association Annual Conference 2013*, Vol. 1. Edited by A. Brown, L. Figueroa, and C. Wolkersdorfer. Golden, CO: International Mine Water Association. pp. 469–474.

Steefel, C.I. 2007. *CrunchFlow User's Manual.* Berkeley, CA: Lawrence Berkeley National Laboratory.

Steefel, C.I., and Lasaga, A.C. 1994. A coupled model for transport of multiple chemical species and kinetic precipitation/dissolution reactions with application to reactive flow in single phase hydrothermal systems. *Am. J. Sci.* 294:529–592.

Steefel, C.I., and MacQuarrie, K.T.B. 1996. Approaches to modeling of reactive transport in porous media. In *Reactive Transport in Porous Media.* Edited by P.C. Lichtner, C.I. Steefel, and E.H. Oelkers. Reviews in Mineralogy, Vol. 34. Washington, DC: Mineralogical Society of America. pp. 85–129.

Steefel, C.I., and Van Cappellen, P. 1998. Special issue: Reactive transport modeling of natural systems (editorial). *J. Hydrol.* 209:1–7.

Steefel, C.I., DePaolo, D.J., and Lichtner, P.C. 2005. Reactive transport modeling: An essential tool and a new research approach for the earth sciences. *Earth Planet. Sci. Letters* 240:539–558.

Steefel, C.I., Appelo, C.A.J., Arora, B., Jacques, D., Kalbacher, T., Kolditz, O., Lagneau, V., Lichtner, P.C., Mayer, K.U., Meeussen, J.C.L., Molins, S., Moulton, D., Shao, H., Šimůnek, J., Spycher, N., Yabusaki, S.B., and Yeh, G.T. 2015. Reactive transport codes for subsurface environmental simulation. *Comput. Geosci.* 19:445–478. doi: 10.1007/s10596-014-9443-x

Stokes, H.N. 1901. *On Pyrite and Marcasite.* USGS Bulletin 186. Washington, DC: U.S. Government Printing Office.

Stollenwerk, K.G. 1994. Geochemical interactions between constituents in acidic groundwater and alluvium in an aquifer near Globe, Arizona. *Appl. Geochem.* 9:353–359.

Thoenen, T., Hummel, W., Berner, U., and Curti, E. 2007. *The PSI/Nagra Chemical Thermodynamic Database 12/07.* Villigen PSI, Switzerland: Paul Scherrer Institut. https://www.psi.ch/les/DatabaseEN/PSI-Bericht%2014-04_final_druckerei.pdf.

Thorstenson, D.C. 1984. *The Concept of Electron Activity and Its Relation to Redox Potentials in Aqueous Geochemical Systems.* Open-File Report 84-072. Denver, CO: U.S. Geological Survey.

URS Corporation. 2006. *Regional Three-Dimensional Groundwater Flow Model of Buckhorn Mountain.* Prepared for Washington Department of Ecology.

USEPA (U.S. Environmental Protection Agency). 2003. *EPA and Hard Rock Mining: A Source Book for Industry in the Northwest and Alaska.* Seattle, WA: USEPA. www.infomine.com/library/publications/docs/USEPA2003.pdf (accessed April 12, 2017).

USEPA (U.S. Environmental Protection Agency). 2007. WhAEM2000, Wellhead Analytic Element Model, Version 3.2.1. Washington, DC: USEPA. https://www.epa.gov/exposure-assessment-models/whaem2000 (accessed April 19, 2017).

USEPA (U.S. Environmental Protection Agency). 2009. *Guidance on the Development, Evaluation, and Application of Environmental Models.* EPA/100/K-09/003. Washington, DC: USEPA.

USGS (U.S. Geological Survey). 2005–2015. *National Field Manual for the Collection of Water-Quality Data: U.S. Geological Survey Techniques of Water-Resources Investigations*, Book 9, Chapters A1–A10. http://pubs.water.usgs.gov/twri9A (accessed April 14, 2017).

USGS (U.S. Geological Survey). 2017. ModelMUSE groundwater software, Version 3.9. http://water.usgs.gov/nrp/gwsoftware/ModelMuse/ModelMuse.html (accessed April 19, 2017).

Usher, B., Strand, R., Strachotta, C., and Jackson, J. 2010. Linking fundamental geochemistry and empirical observations for water quality predictions using GoldSim. In *International Mine Water Association Symposium 2010: Mine Water and Innovative Thinking*. Sydney, NS: Cape Breton University Press. pp. 313–316. www.imwa.info/docs/imwa_2010/IMWA2010_Usher_486.pdf (accessed July 22, 2013).

Van der Lee, J., De Windt, L., Lagneau, V., and Goblet, P. 2003. Module-oriented modeling of reactive transport with HYTEC. *Computat. Geosci.* 29:265–275.

Van der Sloot, H.A., and van Zomeren, A. 2012. Characterisation leaching tests and associated geochemical speciation modelling to assess long term release behaviour from extractive wastes. *Mine Water Environ.* 31(2):92–103.

Van Genuchten, M.Th., Simunek, J., Leij, F.J., and Sejna, M. 2009. RETC for Windows, version 6.02, Code for Quantifying the Hydraulic Functions of Unsaturated Soils. www.pc-progress.com/en/Default.aspx?retc (accessed July 29, 2013).

Vandenberg, J.A., Lauzon, N., Prakash, S., and Salzsauler, K. 2011. Use of water quality models for design and evaluation of pit lakes. In *Mine Pit Lakes, Closure and Management*. Edited by C.D. McCullough. Nedlands, Australia: Australian Centre for Geomechanics, University of Western Australia. pp. 63–80.

Villinski, J.E., O'Day, P.A., Corley, T.L., and Conklin, M.H. 2001. In situ spectroscopic and solution analyses of the reductive dissolution MnO_2 by Fe(II). *Environ. Sci. Technol.* 35:335–376.

Vrugt, J.A., ter Braak, C.J.F., Clark, M.P., Hyman, J.M., and Robinson, B.A. 2008. Treatment of input uncertainty in hydrologic modeling: Doing hydrology backward with Markov chain Monte Carlo simulation. *Water Resour. Res.* 44(12):WR006720.

Wagman, D.D., Evans, W.H., Parker, V.B., Schumm, R.H., Halow, I., and Bailey, S.M. 1982. The NBS tables of chemical thermodynamic properties. *J. Phys. Chem. Ref. Data* 11(Suppl. 2):1–392.

Walter, A.L., Frind, E.O., Blowes, D.W., Ptacek, C.J., and Molson, J.W. 1994. Modelling of multicomponent reactive transport in groundwater, 1. Model development and evaluation. *Water Resour. Res.* 30:3137–3148.

Wang, H.F., and Anderson, M.P. 1982. *Introduction to Groundwater Modeling: Finite Difference and Finite Element Methods*. New York: W.H. Freeman.

Washington Department of Ecology. 2005. *Buckhorn Mountain Supplemental Draft Environmental Impact Statement: Geochemistry, Surface Water, and Groundwater Quality Discipline Report*. October 26. Olympia, WA: Washington Department of Ecology.

Washington Department of Ecology. 2006. Final Supplemental Environmental Impact Statement, Volume I—FSEIS. September 15. Olympia, WA: Washington Department of Ecology.

Washington Department of Ecology. 2007–2013. Fact Sheet for NPDES Permit No. WA0052434. Buckhorn Mountain Mine. https://fortress.wa.gov/ecy/wqreports/public/f?p=110:302:3310075075980434::NO:RP:P302_PERMIT_NUMBER:WA0052434 (see Public Notice Date 02/27/2014: WA0052344_BuckhornFinalFactSheet_02-27-2014.pdf).

Waterloo Hydrogeologic. n.d. Visual MODFLOW Flex software, Version 4.0. https://www.waterloohydrogeologic.com/visual-modflow-flex/ (accessed April 19, 2017).

Weisener, C.G. 2003. Novel spectroscopic techniques to characterize mine waste. In *Environmental Aspects of Mine Wastes*. Edited by J.L. Jambor, D.W. Blowes, and A.I.M. Ritchie. Ottawa, ON: Mineralogical Association of Canada. pp. 181–202.

Wels, C., Shaw, S., and Royle, R. 2000. A case history of intrinsic remediation of reactive tailings seepage for Questa Mine, New Mexico. *ICARD 2000: Proceedings from the Fifth International Conference on Acid Rock Drainage*. Littleton, CO: SME. pp. 441–458.

Wels, C., Lefebvre, R., and Robertson, A.M. 2003. An overview of prediction and control of air flow in acid-generating waste rock dumps. In *Proceedings of the Sixth International Conference on Acid Rock Drainage*. Carlton South, VIC: Australasian Institute of Mining and Metallurgy.

Whetstone Associates. 2010. Baseline Geochemical Characterization Study, Blackfoot Bridge Mine EIS. Prepared for U.S. Department of Interior Bureau of Land Management, Idaho Falls District, Pocatello Field Office.

White, A.F., Delany, J.M., Narasimhan, T.N., and Smith, A. 1984. Groundwater contamination from an inactive uranium Mill tailings pile. 1. Application of a chemical mixing model. *Water Resour. Res.* 20:1743–1752.

Whiting, K.S. 1992. The thermodynamics and geochemistry of arsenic, with application to subsurface waters at the Sharon Steel Superfund site at Midvale, Utah. Master's thesis, Colorado School of Mines, Golden, Colorado.

Whittemore, D.O., and Langmuir, D. 1975. The solubility of ferric oxyhydroxides in natural waters. *Ground Water* 13:360–365.

Williams, R.D., and Diehl, S.F., eds. 2014. *Techniques for Predicting Metal Mining Influenced Water*. Management Technologies for Metal Mining Influenced Water, Vol. 5. Englewood, CO: SME.

Winter, T.C., Harvey, J.W., Franke, O.L., and Alley, W.M. 1998. *Ground Water and Surface Water: A Single Resource*. Circular 1139. Denver, CO: U.S. Geological Survey. http://pubs.usgs.gov/circ/circ1139/.

Wolery, T.J. 1992. *EQ3/6, A software package for geochemical modeling of aqueous systems: Package overview and installation guide* (Version 7.0). Livermore, CA: Lawrence Livermore National Laboratory. http://wipp.energy.gov/library/cra/2009_cra/references/Others%5CWolery_1992_EQ36_A_Software_Package_for_Geochemical_Modeling_of_Aqueous_Systems_ERMS241375.pdf.

Wolery, T.J., and Daveler, S.A. 1992. *A Computer Program for Reaction Path Modeling of Aqueous Geochemical Solutions: Theoretical Manual, User's Guide, and Related Documentation* (Version 7.0). Report UCRL-MA-PT IV. Livermore, CA: Lawrence Livermore National Laboratory.

Wolery, T.J., and Jarek, R.L. 2003. *Software User's Manual: EQ3/6, Version 8.0.* Albuquerque, NM: Sandia National Laboratories. www.wipp.energy.gov/library/cra/CRA-2014/references/authorFiles/Wolery_TJ_and_RL_Jarek.htm.

Wösten, J.H.M., and van Genuchten, M.Th. 1988. Using texture and other soil properties to predict the unsaturated soil hydraulic functions. *Soil Sci. Soc. Am. J.* 52:1762–1770.

Wright, L.T. 1906. Controlling and extinguishing fires in pyritous mines. *Eng. Min. J.* 81:171–172.

Wright, S. 1979. *Stella Programmer's Guide*. https://cdn.preterhuman.net/texts/computing/general/stella.pdf

Wunderly, M.D., Blowes, D.W., Frind, E.O., and Ptacek, C.J. 1996. Sulfide mineral oxidation and subsequent reactive transport of oxidation products in mine tailings impoundments: A numerical model. *Water Resour. Res.* 32:3173–3187.

Xu, T., White, S.P., Pruess, K., and Brimhall, G.H. 2000. Modeling of pyrite oxidation in saturated and unsaturated subsurface flow systems. *Transport Porous Media* 39:25–56.

Xu, T., Sonnenthal, E., Spycher, N., and Pruess, K. 2004. *TOUGHREACT User's Guide: A Simulation Program for Non-Isothermal Multiphase Reactive Geochemical Transport in Variable Saturated Geologic Media*. Report LBNL-55460. Berkeley, CA: Lawrence Berkeley National Laboratory.

Xu, T., Sonnenthal, E., Spycher, N., and Pruess, K. 2006. TOUGHREACT: A simulation program for non-isothermal multiphase reactive geochemical transport in variably saturated geologic media. *Comput. Geosci.* 32(2):145–165.

Xu, T., Sonnenthal, E., Spycher, N., Zhang, G., Zheng, L., and Pruess, K. 2012. TOUGHREACT: A simulation program for subsurface reactive chemical transport under non-isothermal multiphase flow conditions. In *Groundwater Reactive Transport Models*. Edited by F. Zhang, G.T. Yeh, and J.P. Parker. Sharjah, UAE: Bentham Science Publishers. pp. 74–95.

Yabusaki, S.B., Fang, Y., Williams, K.H., Murray, C.J., Ward, A.L., Dayvault, R.D., Waichler, S.R., Newcomer, D.R., Spane, F.A., and Long, P.E. 2011. Variably saturated flow and multicomponent biogeochemical reactive transport modeling of a uranium bioremediation field experiment. *J. Contam. Hydrol.* 126:271–290.

Yeh, G.T., and Tripathi, V.S. 1989. A critical evaluation of recent developments in hydrogeochemical transport models of reactive multichemical components. *Water Resour. Res.* 25:93–108.

Yeh, G.T., Tripathi, V.S., Gwo, J.P., Cheng, H.P., Cheng, J.-R.C., Salvage, K.M., Li, M.H., Fang, Y., Sun, J.T., Zhang, F., and Siegel, M.D. 2012. HYDROGEOCHEM: A coupled model of variably saturated flow, thermal transport, and reactive biogeocheical transport. In *Groundwater Reactive Transport Models*. Edited by F. Zhang, G.T. Yeh, and J.P. Parker. Sharjah, UAE: Bentham Science Publishers. pp. 3–41.

Younger, P.L., and Robins, N.S. 2002. *Mine Water Hydrogeology and Geochemistry*. London Geological Society.

Younger, P.L., Banwart, S.A., and Hedin, R.S. 2002. *Mine Water: Hydrology, Pollution, Remediation*. Dordrecht, The Netherlands: Kluwer Academic Publishers.

Zemaitis, J.F., Jr., Clark, D.M., Rafal, M., and Scrivner, N.C. 1986. *Handbook of Aqueous Electrolyte Thermodynamics: Theory and Application*. New York: Design Institute for Physical Property, American Institute of Chemical Engineers.

Zhang, F., Yeh, G.T., and Parker, J.P., eds. 2012. *Groundwater Reactive Transport Models*. Sharjah, UAE: Bentham Science Publishers.

Zheng, C., and Bennett, G.D. 2002. *Applied Contaminant Transport Modeling*. New York: Wiley-Interscience.

Zhu, C., and Anderson, G.M. 2002. *Environmental Applications of Geochemical Modeling*. Cambridge: Cambridge University Press.

Zhu, C., and Burden, D.S. 2001. Mineralogical compositions of aquifer matrix as necessary initial conditions in reactive contaminant transport models. *J. Contam. Hydrol.* 51:145–161.

Zhu, C., Anderson, G.M., and Burden, D.S. 2002. Natural attenuation reactions at a uranium mill tailings site, western USA. *Ground Water* 40:5–13.

Index

NOTE: *f.* indicates figure; *t.* indicates table.

M

N